Dear Martin,
   All the best

      James Kempf

      [signature]

Wireless Internet Security

Architecture and Protocols

Starting from a foundation in the tools of network architecture development and cryptographic algorithms, this text approaches wireless Internet security from the position of system architecture. The focus is on understanding the system architecture of existing Internet security protocols used widely in wireless Internet systems, and on developing architectural changes to counter new threats.

The book begins with an introduction to the topics of security threats in wireless networks, security services for countering those threats, and the process of defining functional architecture for network systems. Examples of cryptographic algorithms are included, and the author goes on to discuss examples of wireless Internet security systems such as wireless network access control, local IP subnet configuration and address resolution, IP mobility, and location privacy. Each chapter describes the basic network architecture and protocols for the system under consideration, the security threats faced, a functional architecture for the security system mitigating the threats, and the important Internet protocols that implement the architecture. The text is an ideal resource for graduate students of electrical engineering and computer science, as well as for engineers and system architects in the wireless network industry.

James Kempf is a Research Fellow at DoCoMo Labs USA and has been active in systems and software research since he was awarded his Ph.D. in Systems Engineering from the University of Arizona in 1983. Prior to his current position, Dr. Kempf worked at Sun Microsystems for 13 years, where he was involved in a variety of research projects, including, in 1994, a prototype of a SPARC-based tablet computer with early 802.11 supports. His research interests include wireless Internet security, new Internet architectures, and immersive user interfaces for wireless terminals.

Wireless Internet Security

Architecture and Protocols

JAMES KEMPF
DoCoMo Labs USA

CAMBRIDGE
UNIVERSITY PRESS

CAMBRIDGE UNIVERSITY PRESS
Cambridge, New York, Melbourne, Madrid, Cape Town, Singapore, São Paulo, Delhi

Cambridge University Press
The Edinburgh Building, Cambridge CB2 8RU, UK

Published in the United States of America by Cambridge University Press, New York

www.cambridge.org
Information on this title: www.cambridge.org/9780521887830

First published 2008

Printed in the United Kingdom at the University Press, Cambridge

A catalog record for this publication is available from the British Library

Library of Congress Cataloging in Publication data
Kempf, James.
Wireless Internet security : architecture and protocols / James Kempf.
    p.   cm.
Includes bibliographical references and index.
ISBN 978-0-521-88783-0 (hbk. : alk. paper)
1. Wireless Internet – Security measures.   I. Title.
TK5103.4885.K46   2008
005.8 – dc22   2008023549

ISBN 978-0-521-88783-0 hardback

Contents

Preface

Wireless Internet Security: Architecture and Protocols approaches wireless Internet security from the direction of system architecture. A system architecture is essentially a high-level blueprint that guides the detailed design, implementation, and deployment decisions that result in a real, usable system, just like the architectural plans for a building guide its construction. Architectures serve as tools for understanding how to design and evolve a complex information technology system. Architectures are regularly developed by wireless standardization bodies to guide the development of interoperable, standardized protocols on interfaces between equipment provided by multiple vendors, including wireless devices used by consumers. Corporations often provide architectures as guidelines for customers, describing how their products fit together with other equipment to provide solutions for their customers' information technology problems.

In the field of wireless security, the architectural approach has been neglected. This neglect is partially a result of the case-driven nature of network security. Most security systems have been developed in response to specific attacks that surface after the system has been deployed, rather than as a planned part of the initial system development process. Indeed, the original Internet architecture had almost no provisions for security. Internet users were assumed to be members of a co-operative community that would never attempt actions on the Internet harmful to others' interests. This approach is changing slowly, as system designers begin to internalize the disastrous results of grafting security onto a system after a successful attack has compromised the original design.

The other part of the book title, "wireless Internet," is a somewhat broad term that covers two different types of radio links. One type, cellular links, tends to require large and deep wired access networks behind the radio link that utilize specialized protocols to manage the radio link in very detailed and radio protocol-specific ways. Cellular link protocols are quite different from the types of link layer protocols on which the Internet Protocol (IP) has traditionally run. The other type of radio link, noncellular links, does not in principle require deep radio access networks, though some noncellular protocols have introduced them as an optimization for better functioning. These kinds of links are more similar to the traditional types of wired link protocols on which IP runs. In addition, as of this writing, the current generation of cellular systems now widely deployed, 3rd generation systems, includes system interfaces which run traditional telephony protocols that are not from the Internet protocol suite or which run modifications of Internet protocols that are different from other systems. In selecting technical material to cover, I needed to make a decision about where the text should focus, and I chose to emphasize the use of protocols from the Internet protocol suite on noncellular radio links. These types of systems tend to have cleaner architectures and are therefore easier to understand and draw lessons from that can then be applied to more complex systems, such as cellular. Merging the Internet and cellular networks has been a more complex and challenging task than anyone thought it would be when the effort started ten years ago, but the next generation of cellular systems, the All-IP Network or AIPN currently under standardization, should eliminate most of the legacy telephony protocols and come much closer to the goal of having cellular networks fully support the Internet protocol suite.

In this book, Chapter 1 discusses some fundamental issues in security for any network system: security threats, how to assess threats, and basic solutions and services to mitigate threats. Chapter 2 presents the functional architecture approach as a tool for developing an architecture for wireless security systems. In Chapter 3, the cryptographic and other security algorithms important for Internet protocol standards are reviewed. Chapters 1 through 3 present introductory material and can be skipped by those knowledgeable about the topics discussed. Chapter 4 develops an architecture for wireless network access authentication systems and describes two standardized system designs in widespread deployment – AAA server based and hotspot – and the protocols associated with the designs. The material in Chapter 4 illustrates how a security architecture can be instantiated into different system designs depending on the specific implementation and deployment needs. Chapter 5 discusses the security architecture and protocols involved in local IP subnet configuration systems that allow wireless hosts to securely configure an IP address and other information necessary to begin obtaining Internet routing service when they move to a new geographic area. Chapter 6 presents the security architecture and protocols for global IP mobility. Chapter 6 also shows the limits of the architectural approach. Like other information systems technology areas, a good architecture and system design do not help if the implementation introduces bugs. Security flaws can crop up at any point in the design, implementation, and deployment process. Finally, in Chapter 7, a security threat very specific to wireless networks, namely compromise of location privacy, is discussed. Chapter 7 illustrates how a basic architectural change can solve a security problem in a cleaner way, at the expense of deep and possibly expensive changes in implementation and deployment.

Throughout the book, I have attempted to maintain a level of detail for algorithms and protocols sufficient to provide good understanding of how the respective algorithm or protocol works, without overwhelming the reader. Certainly, any implementation effort should consult more comprehensive sources. While an introductory undergraduate course in network security is helpful to provide more depth, consultation of the references for additional information should be sufficient to provide background on the security algorithms. Knowledge of the basic Internet protocol suite, such as TCP and DHCP, and some familiarity with mobility protocols, such as Mobile IP, is assumed. Chapters 4 through 6 review the background on the architecture of the underlying protocols and systems prior to discussing the security architecture and protocols for wireless systems. In Chapter 7, some knowledge of IP routing is required in order to understand how the location privacy architectural enhancements work. Most of these topics are covered in introductory undergraduate networking courses.

Each chapter after the introductory material in Chapters 1 through 3 follows a similar pattern. A particular subsystem important to the functioning of wireless networks is introduced with a review of the architecture and protocols that have been standardized to implement the subsystem. This is followed by a threat analysis and the development of a functional architecture independent of the specific standardized protocols but modeling their functionality. Interfaces are then defined between functional elements, and an overview of the standardized security protocols on those interfaces is presented. Chapter 7 is slightly different, due to the lack of any comprehensive standardized architecture or protocols for location privacy. Instead, the results of a research study in how to modify the IP routing and forwarding architecture are expanded into a functional architecture for location private addressing. The goal of the book is to provide an understanding of the underlying design principles for wireless Internet security systems to students and others seeking to know more about how current systems are designed, as well as a useful guide for designers and system architects modifying existing systems or developing new ones.

Acknowledgements

This book grew out of a tutorial I presented at the Croucher Foundation Advanced Study Institute on Cryptography, December 2004, in Hong Kong, on the current state of wireless Internet security protocols. The meeting gave me an opportunity to meet with other researchers in wireless security and compare notes on the state of the art and where the field was going. I would like to thank the Croucher Foundation organizers, in particular Dr. Frances Yao of City University, Hong Kong, for the opportunity to participate in the meeting. Minoru Etoh, Eisuke Miki and Kazuo Imai, CEOs of Docomo Labs USA and my managers over the three years of intermittent effort required to write this book, were incredibly supportive in what turned out to be a very difficult and demanding task, much more difficult than I envisioned when I started writing. I would like to thank them for that support. I'd like to thank Marcelo Bagnulo and the University Carlos III of Madrid for the opportunity to give a one week seminar in June 2007 on Chapters 1 through 4. The interaction with the seminar attendees helped me refine the material in these chapters. I would also like to thank my dedicated reviewers, Erik Guttman, Cedric Westphal, and Renate Kempf, for their efforts in reviewing the book before it was submitted for publication. Any errors are of course my own but their work has helped immensely to improve presentation, understandability, and technical accuracy. Finally, I would like to thank my colleagues at the Internet Engineering Task Force and the Internet Architecture Board for many years of stimulating and informative discussion on the technical topics surrounding wireless Internet security and Internet standards in wireless and mobile networks.

1 Security basics

The Internet was originally developed with little or no security. As a government-run test bed for academic research, the user community was co-operative and nobody considered the possibility that one user or group of users would undertake operations harmful to others. The commercialization of the Internet in the early to mid 1990s resulted in the rise of the potential for adversarial interactions. These interactions are motivated by various harming concerns: the desire for profit at others' expense without providing any offered value, the need to prove technical prowess by disruption, etc. The introduction of widespread, inexpensive wireless links into the Internet in the late 1990s led to additional opportunities for disruption. Unlike wired links, wireless links know no physical boundaries, so physical security measures that are effective for securing the endpoints where terminals plug into wired networks are ineffective for wireless links. Some initial attempts to secure wireless links had the opposite effect: providing the appearance of security while actually exposing the end user to sophisticated attacks. Subsequently, wireless security has become an important technical topic for research, development, and standardization.

In response to the rise of security problems on the Internet, the technical community has developed a collection of basic technologies for addressing network security. While there are special characteristics of wireless systems that in certain cases distinguish wireless network security from general network security, wireless network security is a subtopic of general network security. Many of the same problems, design approaches, and even protocols that have been developed for wired network security can be applied to wireless network security too. This chapter discusses the background topics that are important in any discussion of network security. Specifically, in this chapter we discuss the importance of a threat analysis to good security architecture, and we review different classes of threats to network security that are important for wireless networks as well. We then review the general classes of security services that are available to help mitigate the threats. These services are each associated with specific cryptographic algorithms, which we review in Chapter 2. Finally, we discuss additional security systems that provide support for the security services. In some cases, these systems are also associated with particular algorithms discussed in Chapter 2. This chapter serves as a foundation for later application specifically to wireless networks.

1.1   Importance of a threat analysis

Network security protocols are necessary on the Internet because some people are motivated to exploit or disrupt communications for financial gain or simply to prove their technical ability to do so. In addition, communications between two parties might sometimes be sensitive or involve money changing hands, in which case both parties to the communication have an interest in security. While these points might seem obvious now, they certainly were not obvious to the original designers of the Internet, since no security was incorporated into the original Internet architecture. Until the Internet was commercialized in the mid 1990s, nobody took security seriously in protocol design, with the exception of government agencies that used the Internet protocol for defense and intelligence purposes and researchers interested specifically in cryptography and other security topics.

Security problems usually result from network protocols or systems that contain opportunities for unauthorized or disruptive activity in their design. An opportunity presented by a particular network protocol or system for an unauthorized party to disrupt, harm, or exploit the network communications of two legitimately communicating parties constitutes a *threat* against the protocol or system. A particular sequence of protocol messages and computations which successfully exploits such an opportunity is an *attack*. Much of network security involves identifying threats, figuring out how attacks can be mounted, and then designing fixes to protocols – or, even better, incorporating security into protocol designs before they are finalized – to thwart attacks.

For network systems in general, two important steps in developing an architecture and designing the protocols are to define the problem and to list the characteristics of an acceptable solution. Without a clear and concise problem statement, it is hard to develop an architecture or design a protocol, because a network system, like any other work of engineering, is a designed object that is meant to address a specific problem. For example, the original design of the Internet architecture solved the problem of how to interconnect many different kinds of incompatible network link types, like Ethernet, ATM, etc. Once the problem is defined, a list of characteristics for an acceptable solution, often called *requirements*, serves to limit the solution space in order to direct design energy toward the most promising architectural solution. Without requirements, much time and energy can be wasted on adding features to the architecture that are marginally useful, or critical features can even be overlooked. Requirements also serve to highlight engineering tradeoffs – where sets of features are in conflict – and therefore where compromises must be made in the design in order to come up with something that really can be implemented and deployed. The equivalent activity for security – identifying the threats and figuring how attacks can be mounted – is called a *threat analysis*.

1.1.1   How to conduct a threat analysis

In most cases, a threat analysis starts from an existing architecture, protocol or system design. Ideally, the threat analysis should begin when the underlying network system architecture is complete but before protocol design has started. Starting prior to that is difficult, because it is hard to spot opportunities for attacks if the basic functions of the underlying system are still unknown. A threat analysis may result in changes to the underlying network system architecture, but changes in the network system architecture prior to protocol design are typically not difficult. Waiting until the protocol design is complete – which was all too often the case for older protocols that were not designed based on a good security architecture – runs the risk of having to go back and make major changes in the system architecture to enable a more secure protocol design or accepting compromises in the security imposed by existing implementations.

A threat analysis is conducted by finding opportunities for disruption or compromise of communication. The following factors in a network architecture, system, or protocol contribute to generating threats:

- An unprotected function in the architecture, protocol, or system design, implementation or deployment that offers a dedicated and knowledgeable opponent an opportunity to attack. An example of such a weakness is a sensitive communication between two parties that is conducted in the clear, so that it can be interpreted by an eavesdropper.
- A weakness in the protocol or system design, implementation, or deployment that allows inadvertent disruption of communications, where the disrupting party is actually not intending to attack. Inadvertent disruption factors are typically not architectural in nature, since they usually arise from unanticipated bugs in a protocol or system design. An example is using a transport protocol without built-in congestion control that does unrestricted retransmission without any backoff. Such a protocol could result in severe congestion if many terminals started transmitting at once, denying service to other applications and terminals on the network.
- Some basic parts of the network infrastructure can be attacked in crude and simple ways that cannot realistically be defended against. For example, an attacker could open the door of a microwave oven in an 802.11b wireless LAN cell, disabling any wireless LAN communications for some radius around the microwave oven because both 802.11b and microwave ovens use approximately the same radio frequency.

Architectural solutions are not always the best way to handle a threat. For example, in the case of an 802.11 microwave oven attack, the defense is to find the microwave oven and close the door. The alternative solution of locking up all the microwave ovens in the building and requiring some kind of credentials check to use them is unrealistic and not really commensurate with the threat. This is an example of how a threat can be handled as part of the network system deployment. If the threat is not architectural in nature, then architectural solutions are obviously not the right way to address it. For example, if an application protocol uses a transport protocol without backoff for retransmission, the solution is to modify the protocol design to include proper backoff.

After threats have been identified, the next step is to generate some realistic assumptions about the nature of the attacker. If the assumptions are too lax, serious threats may be overlooked leading to attacks when the protocol or system is deployed. On the other hand, if the assumptions are too strict, the security solution may be overengineered for the actual threat. Most publicly visible mistakes in assumptions about the attacker tend to be on the lax side, since these tend to result in spectacular and widely published security failures when products are deployed and someone manages to crack the security. Assumptions on the too strict side usually delay a product's deployment, cause cost overruns, or require users to jump through so many unnecessary security hoops that the product fails from a usability standpoint. These failures tend to look less like security failures and more like failures in engineering management and product design.

A standard assumption about the attacker when conducting a threat analysis is that the attacker is able to see all traffic between legitimate parties to the protocol. While this assumption may not be true for most wired networks, it is almost always true for wireless networks. Given that, the next assumption is that the attacker can alter, forge, or replay any message they have intercepted. This allows the attacker to impersonate one of the legitimate parties or otherwise attempt to get the legitimate parties to do what they want. The attacker is also assumed to be able to reroute messages to another party, so that the attacker can team up with others to increase the computational and network power available. Finally, the attacker is assumed to have the ability to compromise cryptographic material used to secure traffic if the cryptographic material is sufficiently old. The safe age depends on the type and strength of the cryptographic material. Assumptions about the identity of the attacker are also important. Many attacks are perpetrated by insiders who are known and authorized users, but who misbehave unintentionally due to compromise of their terminals by viruses or malware or perhaps intentionally due to some unknown motivation. A threat analysis cannot assume that known users will never be a threat.

The amount of knowledge and resources available to the attacker typically determine whether the attacker can exploit a particular opportunity for attack, and therefore which threats should have priority for mitigation. It is never wise to assume that an attack can be deterred by keeping the attacker in ignorance about how a protocol works. Most attackers, if they are motivated to attack at all, are willing to expend the time and energy necessary to understand how to make their attack successful. Such *security by obscurity* is an invitation to attackers to crack the protocol or system, and thereby gain an enhanced reputation in "black hat" (bad guy) circles for their cleverness. On the other hand, increasing the amount of resources necessary to mount an attack – so that a successful attack becomes difficult or impossible to mount with a commonly available set of resources – is a legitimate and often-used method of deterring an attack. As we will see in the next chapter, it is actually the basis of mathematical cryptography. However, since computing power is constantly increasing and new mathematical understanding occasionally causes old cryptographic algorithms to become easily breakable, any defense based on increasing the amount of resources by a finite amount must consider where the boundary for a successful attack lies. Architectures and protocol designs that incorporate flexibility for strengthening cryptographic parameters and algorithms, or increasing the computational power necessary to compromise a system should the boundary be reached are an important way of ensuring that designs keep current.

An important consideration when performing a threat analysis is to clearly identify the value of the threatened activity or the severity of the disruption. If the value of the activity is low or the severity of the disruption is slight, measures to counteract the threat should be similarly lightweight. However, care should be taken when making value judgments in this manner, since sometimes threats that are considered unlikely or minor become more important as a protocol or system is more widely deployed. Sometimes, threat mitigation measures are not intended to remove the possibility of attack entirely, but just to reduce the threat to a level that existed before the protocol or system was developed. Of course, this doesn't help solve the underlying problem in the deployed protocols or systems, but sometimes such mitigation to existing threat levels is the only realistic choice, given implementation and deployment constraints.

The process of conducting a threat analysis is unfortunately very heuristic and not very quantitative. A successful threat analysis is best conducted by donning the mindset of the attacker. The person conducting the analysis needs to ask in what clever and creative ways the particular functioning of the protocol or system can be disrupted. In the rest of the chapter, we will discuss some generic classes of threats and the security services that have evolved to counter them. Looking for these classes of threats is a good starting point when conducting a threat analysis. In Chapter 2, we discuss in more detail how a threat analysis is incorporated into the process of developing a security system architecture.

1.2   Classes of threats

While every network protocol or system has particular characteristics that render it more or less susceptible to attack, a few basic classes of attacks are repeated with various permutations in different circumstances. The basic threat classes apply to wireless networks as well. The basic threat classes are:

- replay threat
- eavesdropping and spoofing
- man-in-the middle (MitM) threat
- denial-of-service (DoS) threat.

Network security architectures, protocols, and systems have evolved to counter attacks based on these threats using various kinds of cryptographic and other security algorithms. In this section, we briefly examine each class of threat.

1.2.1   Replay threat

A *replay attack* occurs when the attacker is able to capture traffic from one party and replay it to another, causing the targeted party to perform actions as if the traffic had been received from a legitimate sender. Replay attacks are often coupled with other attacks, such as man-in-the-middle attacks or denial-of-service attacks. In the first case, the replayed traffic is captured due to the attacker's position as a man in the middle. In the second, the replayed traffic is used to take advantage of a flaw in the protocol design or implementation which makes the protocol vulnerable to denial of service.

1.2.2   Eavesdropping and spoofing

Eavesdropping occurs when an attacker that is not a legitimate party to a conversation manages to obtain the contents of traffic between the legitimate parties. The attacker can somehow listen in on the conversation between the parties and use the information it gains. Eavesdropping is primarily a passive activity; the attacker does not engage in a packet exchange with any of the legitimate parties while eavesdropping. The attacker extracts the information of interest from the overheard packet exchange. However, in order for the attacker to set up so that it can eavesdrop, the attacker may have to perform some kind of active packet exchange with the legitimate parties to the conversation or with other parties.

Spoofing means an attacker poses as a legitimate party for the purpose of tricking other legitimate parties into revealing compromising information, stealing service, or for other illegitimate purposes. One reason an attacker may spoof is to enable eavesdropping. Spoofing is typically an active attack, in which the attacker must exchange packets with the local router or a terminal in order to establish its fake identity. Once the identity is set up, the victim begins the network conversation and the attacker is free to manipulate the victim however they see fit.

1.2.3   Man-in-the-middle threat

Man-in-the-middle (MitM) attacks occur when the attacker manages to position themselves between the legitimate parties to a conversation. The attacker spoofs the opposite legitimate party so that all parties believe they are actually talking to the expected other, legitimate parties. A MitM attack allows the attacker to eavesdrop on the conversation between the parties, or to actively intervene in the conversation to achieve some illegitimate end.

MitM attacks are relatively uncommon in the wired Internet, since there are very few places where an attacker can insert itself between two communicating terminals and remain undetected. For wireless links, however, the situation is quite different. Unless proper security is maintained on wireless last hop links, it can be fairly easy for an attacker to insert itself, depending on the nature of the wireless link layer protocol.

1.2.4   Denial-of-service threat

Denial of service (or DoS) occurs when an attacker attempts through some means to reduce the ability of a network or server to provide service to legitimate users. The nature of such attacks can run from crude to extremely sophisticated. For example, in an 802.11b or g (WiFi) wireless network, a crude DoS attack can be mounted by breaking the safety interlock on a microwave oven, then opening the door and starting up the oven. The radio noise generated by the microwave, which operates on the same frequency as the 802.11b and g wireless protocols, will overwhelm the signal from the access points. The threat from microwave ovens is fairly easy to counter, however, since the attacker and the oven must be physically located near the access point to perpetrate the attack, and therefore can presumably be quickly found. Other types of DoS attacks listed in the following subsections, are harder to detect because the attacker can be remote.

Bombing attacks

A more serious but still crude attack is when the attacker bombards a network or server with packets designed to increase network utilization and thereby decrease throughput. Such an attack is especially effective if the attacker controls a network of machines, called *zombies*, throughout the Internet that have been compromised using viruses or spyware. The attacker can then instruct the machines to target a specific website or other service in order to blackmail the owner or otherwise extract some illegitimate benefit. The zombies allow the attacker to perpetrate the attack without exposing its identity, making the attacker difficult to track down. The only currently known way to handle such distributed denial-of-service attacks (DDoS attacks) is to provision the server or network with enough spare capacity so that some legitimate users can always get service, perhaps at a reduced level, or leave some capacity in reserve to be switched on for such situations.

Protocol bugs and DoS attacks

More sophisticated DoS attacks exploit particular weaknesses in protocol design. For example, consider a client-server protocol that takes requests from initially unknown clients, then replies to authenticate the client and set up a session. If the server maintains any outstanding state between the initial request from the unknown client and subsequent responses, the server can be subject to a storage depletion attack. The attacker continually sends the protocol initiation messages from different IP addresses without actually continuing the protocol. At some point, the server may run out of storage for the state and be unable to respond. The solution to such an attack is to design the protocol so that the server does not maintain any outstanding state from the client until the client has been authenticated. Note that this attack is not really specific to wireless networks.

Redirection attacks

A particular kind of DoS attack, called a redirection attack, is a consideration in the design of wireless protocols. A redirection attack occurs when the attacker sets up a session with a server for a large bandwidth data flow, such as streaming video, then redirects the attack at a victim whose network connection or device does not have the bandwidth to handle the flow volume. The victim's network connection is overwhelmed by the traffic and legitimate service grinds to a halt.

Address spoofing

Finally, another attack that is not specific to wireless networks but easier to perpetrate there and therefore more common on wireless networks is address spoofing. The protocol used by IP networks on the last hop for routing has traditionally not been secure, because wired networks have in the past typically operated in situations where physical security or difficulty of access (as for example in dial-in networks) have made attacks unlikely. This protocol allows a router to map an IP address to a link layer address, so that the router can deliver the packet directly to the terminal's interface card through the link layer. However, because the protocol is not secure, it is possible for an attacker on the same link to claim to own the IP address. The router then ends up sending packets to the attacker instead of to the legitimate owner of the address.

1.3   Classes of security services

With the exception of DoS attacks, security services have been developed to counter the threats discussed in the previous section. Security services have many uses in general network security, and are an important part of wireless network security too. For example, unlike wired networks, in a wireless network, any properly configured device within the broadcast radius of a wireless access point can hear the communication between a wireless device and the wireless access point. Depending on the wireless link protocol, an eavesdropping attacker may be able to easily decode the communication and respond as the victim. If a sender on a wireless link wants to prevent eavesdropping, the messages sent and received over the link must include proof of origin to provide data origin authentication, must be encrypted to provide confidentiality protection, and must be protected against replay to avoid use of a previous message by an adversary. These are the basic security service classes. For DoS attacks, most mitigation measures focus on deployment or network management, with the exception of protocol design measures that limit opportunities for DoS. Since DoS attacks exploit some very deep and fundamental properties of the Internet architecture, they are hard to mitigate with specific system architectural measures. Most DoS attacks are also not specific to wireless networks, so they are not discussed further in the book unless they are related to specific protocol design issues.

1.3.1   Data origin authentication

Data origin authentication is the process by which a receiver of a message is able to prove that the message originated from the reputed sender, and that the contents of the message were not altered en route. Data origin authentication is done for every packet in a protocol conversation if the two parties want to make sure that they are talking to each other and that no packets are modified in transit. Sometimes data origin authentication is called *integrity protection*, emphasizing the second aspect, proving that the message was not altered, rather than the first, proving that the message originated from the reputed sender.

Data origin authentication requires cryptographic techniques in order to construct the proof of origin. The cryptographic techniques require that the two parties to the conversation possess cryptographic material that allows one party to construct a proof of authenticity and allows the other party to check it. The cryptographic material is usually in the form of a cryptographic key or keys. Later in this chapter, we discuss keys and their distribution.

The sender constructs the proof by taking some kind of summary of the message, usually packet by packet, and performing a cryptographic operation on the summary that only the possessor of the sender's key could perform. The summary is a short number of bytes that uniquely identify the message, and which the receiver can calculate directly from the message too. The summary must be long enough that an attacker cannot easily construct it by guessing or otherwise cheating. The receiver verifies the cryptographic proof using a matching key. When the following two conditions hold the receiver can deduce that the message was not modified in transit and did, in fact, originate with the sender:

- The summary constructed by the receiver matches the summary constructed by the sender.
- The receiver is able to verify a proof that only the sender can construct.

1.3.2   Confidentiality protection

When people think about network security, confidentiality protection is often the first thing that comes to mind. *Confidentiality protection* allows the sender of a message to know that only a designated receiver is able to read the contents of the message, and that the message is unreadable to unintended eavesdroppers. Confidentiality protection is usually achieved by encrypting the message. Encryption uses some cryptographic material and a cryptographic algorithm to convert a plain text message into a cipher text. The cipher text message is in theory not decipherable unless the receiver has matching cryptographic material and knows the cryptographic algorithm by which the message was encrypted. To an outside observer without the matching cryptographic material, the cipher text looks like randomly generated bits. The receiver uses the matching cryptographic material and cryptographic algorithm to decrypt the message into the original plain text.

Many different kinds of encryption algorithms having various properties are available, and in Chapter 2 we discuss two representative samples that are in wide use and provide good protection in general. Encryption, like data origin authentication, requires both sides to have a collection of cryptographic material. The kind of encryption algorithm used in a particular wireless security system design is often determined by the kind of key distribution protocol available. The processing power available for cryptography is also an important consideration when selecting a cryptographic algorithm, since each packet requires processing. Additionally, while there are many encryption algorithms that provide good protection properties when used correctly, some algorithms have flaws or weaknesses that require consideration when including them into a design. Finally, a wireless security system should never use the same cryptographic material for encryption and data origin authentication, even if the same cryptographic algorithm is used for both. Using different cryptographic material ensures that if an attacker somehow manages to break encryption, for example, data origin authentication is still protected until the attacker has a chance to break that.

1.3.3   Replay protection

Replaying previous messages captured during a legitimate transaction is another possible attack that can be perpetrated. The replayed messages can clog up the victim's processing, thereby denying service to legitimate correspondents, or they can be intended to elicit the same response that the victim provided when the message was originally received, but this time to the attacker. In either case, replay protection is an important part of general network security protocols, and is also needed in wireless security protocols.

Replay protection is usually achieved by having the sender include in each message a sequentially increasing sequence number. The receiver then validates that the sequence number corresponds to an already-received packet. If the sequence number was already received, the receiver discards the packet. In order to avoid spoofing, the sequence number must be covered by data origin authentication. Otherwise, an attacker could modify the sequence number on a legitimate packet in order to cause the replay protection mechanism at the receiver to reject the packet. Many network protocols include a sequence number so that requests and replies can be matched. Therefore, providing secure replay protection often requires little more than that the protocol include data origin authentication in addition, to protect the sequence number.

1.4   Supporting security systems

To perform their function, the cryptographic algorithms providing the security services discussed above require cryptographic material. Some means is required to securely provision and manage cryptographic material. The collection of cryptographic material, credentials, and identifiers for these items shared between two sides, together with the associated cryptographic algorithms to which the provisioned material apply, is called a *security association*. The most important part of the provisioned cryptographic material in a security association is typically the key used for the cryptographic algorithm, so key management is the basis of security association management. The next subsection discusses key management.

A prerequisite for establishing a security association is that both sides of the transaction can verify an authenticated identity of the other. In addition, most network operators require some method whereby the identity of a network node wishing to obtain network service is verified, and the rights for particular services are authorized. If the user is charged for service, network usage and their cost must be recorded. The algorithms, protocols, and systems that implement identity management provide an important support role for the basic security services, and are the topic of this section. The subsection following the next discusses identity management.

1.4.1   Key management

Security processes such as data origin authentication and encryption require that both sides of a network conversation share cryptographic material, or *keys*, allowing them to perform specific cryptographic algorithms in common. Arranging for both sides to have the right keys prior to the need for cryptography is key exchange or *key management*. Designing architectures and protocols for the secure provisioning of keys and management of keys over time is one of the most difficult and complex parts of designing good wireless network security systems.

The security properties of the key management system often depend on what type of cryptographic algorithm is used. Cryptographic algorithms come in two basic types:

- shared, secret, or symmetric key algorithms
- public or asymmetric key algorithms

We discuss shared key and public key cryptographic algorithms in more detail in Chapter 2, the rest of this section provides important background material on designing shared and public key management systems.

Key management for shared keys

In shared key algorithms, both sides to the conversation share the same cryptographic key. The key must be kept secret from all other parties. If the key is ever revealed, the discovering party will have the ability to perform the same cryptographic operations as the two legitimate parties. This could allow the discovering party to masquerade as one of the two legitimate parties, or to decrypt encrypted messages between the two legitimate parties.

The need for keeping the key secret requires either that:

- one side of the conversation (typically a key provisioning server) generates a shared key and securely sends it to the other, or
- both sides of the conversation deduce the shared key independently using an algorithm without exchanging any confidential material over the network.

If the key is sent from one party to the other, the provisioning protocol must be properly designed. This includes data origin authentication, replay protection, and – most importantly – encryption. A terminal being provisioned must be able to verify the identity of the provisioning entity and the key must be protected from eavesdroppers while it is in transit. Transport security on the shared key can be accomplished by using either another symmetric key shared between the two sides, or an asymmetric key.

If both sides deduce the key independently, the algorithmic deduction can take one of two forms:

- The two sides exchange nonconfidential material over the network then deduce the shared key algorithmically without reference to any preshared secret material, using a public key-like algorithm such as Diffie–Hellman.
- The two sides deduce the key algorithmically from some preshared secret material, with possibly some nonconfidential material exchanged over the network for freshness.

The first method is really a subcategory of public/private key management, because the algorithms used are public key algorithms. In Chapter 2, we discuss the Diffie–Hellman key exchange algorithm in more detail.

For the second method, the wireless terminal must be provisioned with the preshared secret in some out-of-band fashion (e.g. typing in some numbers, through a secure hardware chip, etc.) prior to its first network access. The secret is shared with the other party to the key generation. This could be a provisioning server in the network or even another terminal, depending on the application. When the wireless terminal contacts the other party for key generation, the other party and the wireless terminal utilize the preshared secret to generate a *master key*. The key generation algorithm typically requires the combination of the preshared secret with some additional nonconfidential material provided by the wireless terminal, for example, a randomly generated number, exchanged during the key generation transaction. The nonconfidential material provides key freshness. The exact details of master key generation depend on the actual protocol or standard specification. The important point is that both sides independently derive the master key using the same values, one of which is the preshared secret but others of which may be publicly exposed. The master key itself is never actually exchanged between the wireless terminal and the other party, because each side derives it independently. In some protocols, a further step is required in which the other party, usually a key provisioning server in the network, securely conveys the master key to a third party that will ultimately be conducting the cryptographic operations with the wireless terminal. In this case, the key provisioning server and the third party must share a security association specifically for protecting key distribution.

The master key is then used to derive *session keys* for use in various cryptographic operations. The session keys are derived in the same way as the master key: both sides independently combine the master key and some publicly accessible material exchanged between them in a specified, algorithmic fashion. For example, a wireless terminal and a network authentication server may generate a session key for authenticating traffic over the wireless link, a separate session key for encrypting traffic over the link, and yet another session key for securing handover signaling between one wireless access point and another. Session keys typically have a limited lifetime and must be periodically regenerated, to reduce the amount of time they are exposed to an adversary that could compromise them. The regeneration procedure is an important part of the key management algorithm. The master key itself and the preshared secret should never be used directly for cryptographic operations on the network. If a master key or preshared secret is used and somehow an adversary compromises the key, then all derived keys are put at risk too.

Key management for public keys

Public or asymmetric key algorithms do not require the exchange of confidential material or the prior provisioning of both sides with a preshared secret. Instead, each participant in the cryptographic algorithm generates a pair of keys. One key, called the *private key*, is not disclosed to any other party. The other key, called the *public key* (from which this class of algorithms takes its name) is not confidential and can be sent over unencrypted connections to other parties. For most public key algorithms, the public and private keys are calculated algorithmically using random numbers generated autonomously by the owner from a good pseudorandom number generator. The random numbers are then operated upon by the key generation algorithm for a specific public key cryptographic scheme to generate the public and private keys. The random numbers ensure that the private key is mathematically difficult to guess. The key generation algorithm establishes a mathematical connection between the private key and the public key that allows the cryptographic algorithm to work. Public key algorithms are sometimes called asymmetric algorithms because the need for confidentiality on the keys is asymmetric. The public key can be exposed but the private key cannot, unlike shared key algorithms where the confidentiality requirement is symmetric: the shared key must be kept confidential by both sides.

The two keys are used for different security services in different ways. For data origin authentication, the owner of the key pair generates a *digital signature* on data using the private key that allows the receiver to verify the origin and integrity of the data using the public key. For confidentiality protection, the sender of a confidential message to the owner of the key pair uses the public key to encrypt data that allows the owner of the key pair to decrypt the data using the private key thereby protecting it in transit. As mentioned above, in addition to the cryptographic algorithms for data origin authentication and confidentiality protection, public key algorithms also require a key generation algorithm to generate the public key from the private key.

Principles of secure key management protocols

In general, an existing key management protocol having the right characteristics for the application at hand should always be preferred to developing a new key management protocol from scratch. Security protocols usually get better over time because the bugs in them are found and fixed as more and more applications use them. So older protocols – provided they are not so poorly designed as to be in effect insecure – are usually better understood and therefore better to reuse. Of course, to reuse an existing protocol, the assumptions and constraints on the protocol must be carefully noted and not violated; otherwise, a secure protocol can easily be converted into an insecure one.

If an existing protocol is not a good match for a particular system, a new protocol is required. The following principles, discussed in more detail in RFC 4962 (RFC 4962, 2007), have proven successful in mitigating threats in practice and should be kept in mind if a new protocol is developed. These principles are primarily of relevance to key management protocols that provision or derive shared keys:

- Confidentiality protection, replay detection, and authentication are required for key distribution or exchange protocols over the network. Keys are confidential material, and therefore proper security protection is required. In order to prevent spoofing, both sides in the key exchange must be mutually authenticated to ensure that they fully know and trust each other. Finally, replay protection is required to avoid an attacker sending an old session key obtained by snooping a prior exchange, and thereby disrupting an ongoing session.
- The cryptographic algorithm used in a security protocol should be negotiable. The security of cryptographic algorithms is not fixed, and often depends on the processing

power available to an adversary. Since processing power is increasing, the cryptographic algorithms used in a security protocol should be negotiable. This allows parties in the exchange to use the most secure algorithm consistent with their hardware processing power and software implementation availability. In addition, negotiations for selecting a cryptographic algorithm must be performed between authenticated entities, and the messages must be covered by data origin authentication. This prevents an attacker from spoofing one side of the conversation into accepting a weak cryptographic algorithm that the attacker is able to compromise.

- Keys need to be kept strong and fresh. Key freshness means that keys are generated whenever a new session is started, and periodically renewed. A key must be generated specifically for the use that is intended, and the material that goes into calculating the key must be new. In addition, there must be no dependency between keys such that disclosure of a previous key compromises keys that are generated later. Key strength, which is usually a function of the number of bits in the key, must be high enough that the probability of a guessing attack or other compromise is very low. Since the limits of key compromise are changing all the time as computation power increases, protocol developers must be aware of the state of the art in cryptanalysis with respect to key length in order to make wise choices.

- A key in a shared key security association is confidential material, and therefore it should not be divulged, even intentionally, to an entity that doesn't need to know the key. An "entity" here means either a software module on the same node for which the key was derived or another network entity entirely. An entity has access to a key if it has access to all the cryptographic material needed to derive the key. The concept of a *cryptographic boundary* is useful in limiting key access. A cryptographic boundary is a topological scope within which the key is known, but outside of which it is kept secret. A cryptographic boundary encompassing a secure hardware chip is more secure than a software module in the operating system kernel. Similarly, a cryptographic boundary encompassing a single node and the associated server is more secure than one consisting of the server and several other network entities like wireless access points. The smaller the cryptographic boundary, the easier it is to limit potential compromises, and to detect compromises when they occur.

- Authorization checking is required, in addition to authentication. This prevents a terminal that can be authenticated from claiming a higher privilege or more services than it is entitled to. When more than one network entity is involved in the protocol, all must agree on the authorization for the requesting terminal.

- Damage from key compromises should be limited. The compromise of a key is a serious problem, and, although well-designed security algorithms can prevent compromise from passive or active eavesdroppers, compromises in other ways that do not involve an attacker just having access to network traffic are possible. For example, an attacker can get access to a key by stealing the physical hardware device and extracting the key from it. Propagation resistance has many implications, but one is that authenticating entities should never share secret material, and new keys should be derived every time a terminal moves from one authenticating entity to another.

- A unique context for keys should be established consisting of the following information:
  - a unique name;
  - the way in which the key is expected to be used which includes not only the cryptographic operation (for example, data origin authentication) but also the specific application (for example, securing the last hop at the wireless link layer);
  - other parties that are expected to have access to the key;
  - the expected lifetime of the key.

All parties with access to a shared key are expected to agree about the context in which the key is to be used. Each context should have a unique key, in order to reduce the risk that the compromise of one key affects more than one application.

1.4.2   Identity management

Identity management and key management are intimately tied together. A prerequisite for secure key provisioning is a proof procedure whereby both sides to the transaction can verify the identity of the other side. Network operators also require the ability to verify the identity of wireless terminals requesting network access. Key provisioning often accompanies identity verification during network access, since once both sides have verified each other, any keys generated from the transaction are tied to a verified identity. The subsections below discuss identity management for public keys and authentication, authorization, and accounting for identity management during network access.

Public key certificates

Although there is no requirement in public key systems that a public key is kept secret, most applications of public key algorithms require a method allowing a party receiving a public key from another party to cryptographically verify the identity of the party sending the public key. If verification is lacking, it is possible for an attacker to claim the identity of a legitimate party and conduct a transaction with a victim that the victim thinks is authenticated but which in reality is the attacker spoofing the identity of another party. A common way of providing identity verification for the public key is a *public key certificate*. A public key certificate is a collection of structured data containing the owner's public key, information verifiably identifying the owner of the key pair, an indication of the rights of the owner to utilize the public key for various applications, and the expiration date of the certificate. The certificate is signed by an entity, known as a *certificate authority*, whose identity is available, known, and trusted, by a broad variety of other nodes that might want to obtain the verified public key of the public key owner.

The certificate authority in effect vouches for the identity of the public key owner. Naturally, for the public key owner to obtain the certificate authority's signature on the owner's certificate, the public key owner must prove their identity to the certificate authority. The process is similar to obtaining a notarized document. The owner of the document goes to the notary with some kind of proof of identity, like a birth certificate, passport, etc. The owner signs the document in front of the notary. The notary verifies the proof of identity, stamps the document, and signs it. The owner of the document then uses the document to perform a financial or legal transaction of some sort that requires third party proof of identity.

For public key systems, the receiver of a public key certificate verifies the identity of the public key owner by verifying the certificate. The receiver uses the public key of the certificate authority to verify the digital signature on the certificate using a public key authentication algorithm, just as it would with any other data requiring data origin authentication. If the signature matches, the receiver knows that it can trust the information in the certificate, including the public key and identity information. The public key is then said to be *certified*. With a certified public key, a correspondent can trust the identity of the public key owner. This description covers the basics of public key identity management. Chapter 3 discusses a few additional aspects of public key systems that provide more deployment flexibility and security.

Authentication, authorization, and accounting

Owners of wireless networks often want to limit network access to users with whom they have some kind of business relationship. For example, when a business deploys an enterprise wireless LAN network, the business wants to restrict access only to employees. Similarly, access to public access wireless networks such as wireless LAN hot spots or cellular networks is typically restricted to customers who have an account with the network operator or can provide a credit card number for billing. Unlike wired networks, access to wireless networks often does not require that the user have physical access to a particular building or room, so the owners of the network cannot simply impose restrictions on who enters the facilities where the network access is provided in order to restrict who can use the network. The radio signals that carry wireless data often overflow into areas where the owner of the network does not control physical access.

To maintain this kind of control, wireless devices are required to undergo a series of transactions prior to allowing the device full Internet protocol data routablity with the world beyond the immediate wireless link. These transactions consist of the following three operations:

- The device is *authenticated* by requiring it to provide some irrefutable indication of the user's identity and right to use the network.
- The *authorization* of the user for network access and other services is checked.
- If access to the network and other services requires the user to pay, the network sets up *accounting* so that usage of the services can be monitored.

The architecture, protocols, and systems that provide these three functions are often lumped together as authentication, authorization, and accounting or *AAA* (pronounced "triple A"). Together, these three functions provide identity management for network access.

Data origin authentication is often confused with the authentication of the first "A" in AAA, but the two are somewhat different. Authentication in the AAA sense is a matter of proving that a particular user has an account with the owner of the network. It is usually done only once (or potentially once per handover in a wireless network) when the device boots up into the network for the first time. Data origin authentication, on the other hand, is done for every packet exchanged between the parties.

Data origin authentication and AAA authentication do touch at a couple of points. First, any good AAA protocol requires that the messages between the terminal and the AAA server during an AAA session have data origin authentication, so that the two parties to the AAA transaction can have confidence that the messages originated with the reputed sender and were not modified in transit. Otherwise, the AAA server might end up granting access to a device that it thought was authorized, but actually was an attacker, or the device might end up revealing information to a bogus AAA server. Second, after a device has been granted network access, the device and the network often undergo a key management phase, in which both sides configure keys to perform further data origin authentication over the wireless link for the device's traffic. This ensures that only properly authenticated devices can send packets over the wireless link. Since, as mentioned above wireless links tend to be considerably less physically secure than wired links, data origin authentication is often required on the last hop wireless link to prevent unauthorized access even after network access authorization is received by the user.

Many different techniques are used for authenticating and authorizing network access. For authentication, one of the most popular and widespread (but unfortunately least secure) is the username/password. Supplicants wanting network access prove that they are legitimate account holders by typing in a publicly known username and a secret password known only to them. The problem with this system is that people typically choose passwords that are easy to remember but also easy to guess, a characteristic that is said to be low entropy because the passwords are not randomly chosen. Such passwords are subject to simple automated attacks. An example is a dictionary attack where the attacker iterates through a dictionary of commonly chosen passwords until they achieve success. People also tend to reveal their passwords, often for very flimsy reasons. A safer technique is a one-time password, usually supplied by a key card. The password is only valid for a single network access authentication, and is usually generated using a keyed hash function, where the key is shared with the AAA server granting access to the network. The drawback of key cards is that they sometimes break and are easy to lose.

Authorization typically follows directly from authentication; that is, if the supplicant wanting network access proves that it is a legitimate account holder, the supplicant is granted network access. Some deployments may include a service profile in the AAA server, where services to which the user has subscribed other than simple network access are recorded. The initial AAA transaction provides an opportunity to authorize the supplicant for additional services, though exactly how provision of these services is enforced may vary widely. Accounting is also set up at the time the authentication for network access is done. The accounting activates mechanisms in the access network that generate records recording how many packets the user has used if billing is done on a per packet basis or how long the user is connected if billing is done on a per minute or per hour basis. Since many ISPs today provided unlimited service for a monthly fee, accounting for simple network access may be unnecessary, although it may be important for other services.

Many of the AAA protocols and systems in use today for wireless network access had their origin in dialup telephone access to the Internet. The basic problems involved in dialup telephone access of verifying a device or user, checking authorization for network access, and setting up accounting for service billing are superficially similar for wireless networks, which is what led engineers to adopt the same kinds of protocols for wireless deployments. The theory was that since the dialup AAA protocols were widely deployed, it would be much easier and cheaper to leverage off that deployment – the AAA servers and protocols – when setting up wireless networks, rather than deploying a whole new infrastructure for wireless networks. Unfortunately, this theory ignored a couple of key differences between dialup systems and wireless systems. These differences have caused no end of problems and kept engineers busy inventing modifications of AAA protocols to make them work better for wireless networks. In the end, it is debatable whether the strategy of modifying dialup AAA protocols for wireless access really resulted in any significant savings in deployment effort, but the wireless AAA protocols are becoming increasingly widespread and therefore are of importance.

One difference is that in a dialup network, the last hop link between the dialup modem at the user's premises and the IP network can, for all intents and purposes, be considered secure. Once a signal enters the wireline telephone system, it is extremely difficult for an unauthorized device or person to obtain access to the signal. This is not due to any particular combination of technological security features; but rather, is a result of two characteristics of the circuit-switched telephone network:

- The protocols used in the circuit-switched telephone network are not widely known.
- The network itself is designed in a way that makes it difficult to obtain access to an identifiable end-to-end data stream without accessing one of the switches through which the data stream runs.

Since telephone companies tightly restrict who has access to the large switches that run the system, nobody can get access to a call unless they know how and are authorized. In effect, this is a combination of "security by obscurity" and tight control over people who work for the telephone company – not the most modern way to provide security but generally effective given the technology of circuit-switched telephony.

The other major difference between dialup systems and wireless systems is that wireless users rarely stay put. Some wireless users are more nomadic. They move to a particular place, sit down, then work for a while using their wireless device without moving. When they need to move, they usually close up the device and restart their session in a new place. Laptop users are an example. Other wireless users actually use their devices while in motion. Cell phones are an example. In either case, the wireless device may be required to handover from one wireless access point to another, either after restarting or while actual data transmissions are occurring. Dialup users typically never change their point of connection after a particular session has started, and often the same point of connection is used every time a new session is started. Even for non-Internet wireless networks such as the circuit-switched cellular telephone protocols used in the second generation and third generation (2G and 3G) digital voice networks, the wireless medium is inherently insecure because the network operator cannot restrict who has access to the wireless medium, and users move about when making calls.

As a result of this historical legacy, the protocols between the terminal and network adopted into wireless network architectures from dialup AAA initially had little or no security support, and no ability to easily move a fully authenticated and authorized device between one point of connection and another. This had to be modified for wireless Internet access.

The rest of this book does not talk much about the third "A", accounting. That is not because the process of recording network usage in order to collect money is not important. The reason is that basic accounting is not a security function and does not involve any security protocols. Accounting also tends to be more dependent on the particular application, and the business model for the organization owning the network. Accounting for prepaid services is done differently than for services that are billed by the hour, for example. Many private networks, such as corporate networks also do not bill for service and do not need accounting.

1.5   Summary

In this chapter, we have discussed the basic nature of security for wireless Internet systems. An important start to providing security for any network system is to assess the threats against the system. The threats serve as a kind of collection of requirements that the security architecture and protocols need to counter. Some basic classes of threats were described. When a threat analysis is completed and the attacks are understood, the basic security services needed to counter the threats can be determined. The three common security services are data origin authentication, confidentiality protection, and replay protection. Data origin authentication ensures that both parties in a network transaction can verify that the data originated with the expected party and that the integrity of the data was maintained in transit. Confidentiality protection prevents eavesdroppers from listening in on a network transaction. Replay protection ensures that an eavesdropper cannot confuse the correspondents by sending old messages to look like new. These are the basic security services that serve as building blocks of wireless network security architectures.

The basic security services require additional system support for setting up a security association containing cryptographic material, credentials, and other state shared between parties and important to operation of the basic services algorithms. Since security protocols usually require both sides of a secure conversation to possess some kind of cryptographic material, secure and effective key management is an important component of wireless network security. Key provisioning requires identity management to ensure that a provisioned key is tied to a verified identity. Permission to enter a network also requires identity verification, to ensure that the wireless terminal and its user are allowed to use the network.

Finally, DoS attacks are a separate kind of threat that can take a wide variety of forms, most of which can only be countered by deployment measures. Most DoS attacks take advantage of deep and fundamental properties of the Internet architecture, and therefore are difficult to deter with architectural solutions at the subsystem level. DoS attacks on specific network protocols, however, can be countered by ensuring the protocol designs do not incorporate bugs that enable such attacks. DoS attacks of the latter sort are discussed in the following chapters where appropriate for wireless security, but the general topic of DoS attacks requires a larger discussion than is appropriate for this text, since they are not unique to wireless systems.

2 Network system architecture basics

Wireless network operators and end users need the ability to utilize equipment from different vendors in their networks and in customer-accessible devices. Left to themselves, vendors of network equipment and of end-user access devices such as wireless terminals tend to produce equipment that is slightly different in various ways, hindering the ability of their customers to build multi-vendor networks from interoperable equipment pieces. The key to ensuring interoperability is to have a standardized system design with clearly specified interfaces between the various network devices and well-designed, standardized protocols on the interfaces. The process of systematically identifying requirements and functionality and mapping that into network entities, interfaces, and standardized protocols is the key to ensuring a design that meets real-world needs and in which the pieces work together well. This requirement is generally true for network systems, but it also applies specifically to security systems.

While standardization is the key to ensuring interoperability in complex multi-vendor systems, system architectures are the principal tool for guiding the design, implementation, and deployment process. In this chapter, we examine the topic of network system architecture. In the next section, we discuss the role of architecture in system standardization in more detail. Following that, we describe a particular approach to developing a system architecture, the functional architectural approach, that is used in some wireless network standardization processes. We use this approach throughout the book to analyze existing wireless security system architectures, and ultimately in Chapter 7 to add new security architectural enhancements to existing IP systems. To illustrate how network system architectures are developed, we present a simple example of a wireless network system architecture, a key fob used to remotely open a locked car. We then specifically examine how the functional architecture approach works for security systems by developing a functional architecture that provides security for the key fob.

2.1 The role of architecture in system standardization

Most large wireless network systems are developed as part of a standardization process involving multiple vendors and network operations. Since the components of such systems are often manufactured by many vendors and the systems are deployed by many network operators, standardization ensures interoperability between equipment from different vendors and deployments by different operators. The first step in developing a new or enhanced standardized wireless network system is to define a system architecture. The term "architecture" is used in a variety of ways to characterize the initial output of the design process. *Webster's Dictionary* defines architecture as (among other things) "a style or method of design or construction." The approach to architecture for a large-scale wireless network system depends on the process traditionally followed by the standardization body.

Most wireless standardization groups have their heritage in the traditional circuit-switched telephony architecture that was common prior to the widespread adoption of the Internet. These groups adhere to a rigorous system development process, in which formal requirements development is followed by an architectural development phase centered on meeting the requirements. The architectural development phase results in the definition of network boxes with interfaces on which the functions of interoperable protocols are specified. Protocol selection or design follows, after which a formal regression analysis is performed to ensure that the resulting system meets the initially defined requirements. The boxes and protocols are then implemented by vendors as products.

While such a rigorous design process ensures accountability and fidelity with the original requirements, it is often inflexible and unable to account for changes in requirements that come up later in the design process. The process is similar to a waterfall in which the requirements, architecture, protocol design, and implementation fall out of the standardization process like water pouring over a waterfall. Incremental modifications are inhibited, since they are not accommodated by a waterfall development model. The strong emphasis on using the requirements to rigidly structure the architecture often results in pressure by various participants in the standardization process to "game" the requirements, to ensure some advantage for their business or technical positions against their competitors. As a result, the accountability and objectivity that the process was originally intended to provide is usually considerably weakened; most of the important decisions are based on the same kinds of subjective criteria that are behind group decision making in other areas of human endeavor where interests of various parties conflict.

On the other hand, the group responsible for standardizing Internet protocols, the Internet Engineering Task Force (IETF), has traditionally resisted formal architectural definitions on the Internet as a large-scale system. The rationale for this is that, for the Internet as a whole, any attempt to define a detailed architecture would constrain the development of new protocols and new applications too tightly. One of the key factors in allowing the Internet to flexibly accommodate a new generation of applications every five to ten years is the lack of a rigidly fixed architecture overall. Consequently, discussions of architecture at the level of the Internet as a whole are typically confined to a loose collection of design principles, such as those in RFC 1958 (RFC 1958), or adherence to the modified form of the OSI protocol stack (Layer 2, Network, Transport, Application) which informs the design of the IP network stack (Wikipedia, 2008a). As a result, when wireless links became available in the late 1990s, there was no global system architecture for the Internet to guide standardization of interfaces and protocols for wireless networks.

As the Internet has become more complex, however, architectural definitions of well-defined subsystems have become necessary to guide protocol development and ensure interoperability with other subsystems. A good compromise process for defining system requirements that trades off rigor against flexibility has been developed by the IETF. A set of lightweight requirements, called "goals," is developed for the system. The goals are typically as quantitative as possible, but if it is difficult or impossible to assign numbers to what the protocol is supposed to do, a qualitative description is acceptable. The primary distinction between goals and requirements is that there is no intent to regress the final protocol design back onto the goals after the protocol design is complete. The goals are meant to be a set of flexible design guidelines. The same kinds of subjective, non-technical criteria that arise when developing formalized requirements also arise when developing goals. The difference is that because the intent of goals is not to rigidly structure the system/protocol design process, there is more room for flexibility during the design. After the goals are complete, an architecture is developed for the system. The architecture is often called a "framework," and includes descriptions of the major network entities and how they interact at a high level. The protocol design on interfaces between network entities then follows. Not every IETF protocol design follows this process; however, it is often used when new system components are introduced.

2.2 The functional architecture approach

While the frameworks developed during IETF protocol design are good at defining where interfaces between distributed network components need interoperable protocol design, such frameworks are often not very specific about what the different network entities do and what functions the protocol should perform. A functional architecture approach more accurately characterizes these points. The functional architectural approach is more formalized than the framework approach, while, at the same time, maintaining flexibility through the goals. Given a set of goals for a protocol or network system, the functional architecture approach for developing a new subsystem architecture from scratch consists of the following sequence of steps:

1. Using the requirements or goals, define a set of functions which the new subsystem must implement in order to achieve the goals.
2. Group the functions into a set of functional entities with communicating interfaces.
3. Decide which functional entities will be implemented together on a single network device and group these together; communication between the functions on the same device is handled through programmatic interfaces.
4. Define the interfaces between distributed functional entities where protocol design is required.
5. Decide which interfaces are open and require standardized protocols for interoperability purposes and which interfaces are closed and are therefore candidates for vendor-specific protocols.
6. Define what functional actions are communicated across the interfaces and what parameters are required by the functions and what results are returned.

At the conclusion of this process, the design team should have a list of network interfaces on which standardized, interoperable protocol designs are required, a list of closed interfaces (which may be empty) where vendor-specific protocols are needed, and a list of programmatic interfaces between software modules that implement functional entities residing on the same network device. In addition, the list of functional elements and their parameters that need to communicate across a network interface provides a starting point for defining what information needs to be communicated, and therefore what the protocol must do.

As a practical matter, most design exercises these days involve retrofitting new functionality onto existing subsystems with deployed equipment. Backward compatibility is an important constraint, since it ensures interoperability with existing network equipment and thereby reduces the cost of introducing the upgrade. In that case, the list of steps is slightly modified:

1. Using the requirements or goals, define a set of functions which the functionality must implement in order to achieve the goals.
2. Identify which existing network subsystems and which network entities should host the new functions.
3. Group the functions into a set of functional entities and map these onto the existing network entities.
4. Define new communicating interfaces between the new functional entities, or specify how existing interfaces need to be modified to accommodate the new functions. If the interfaces are on the same network entity, then the interfaces are programmatic.
5. Decide which new interfaces are open and require standardized protocols for interoperability purposes and which interfaces are closed and are therefore candidates for vendor-specific protocols.
6. Define what functional actions are communicated across the interfaces and what parameters are required by the functions and what results are returned.

Since most of the wireless security subsystems discussed later in the book were developed as modifications to existing Internet subsystems, we follow this sequence for the examples in Chapters 4 through 7.

A critical point to keep in mind when developing a functional architecture is to avoid committing to a specific protocol solution too early in the design process. Engineers like to think concretely, so there is often a temptation to include protocol solutions as functions rather than wait until the functional architecture is complete before beginning the protocol design. Usually it is possible to tell when a protocol solution is being proposed if someone starts talking about a function as a modification of an existing protocol, about what kind of transport protocol will be used, or about how specific messages will be encoded in protocol data units on the wire prior to the completion of step 1. Of course, if the functional architecture development is for an existing system, then existing protocols constrain the design, but these constraints should not be propagated too far nor too early into the new architectural pieces. Keeping focus on the goals and functional architecture during the initial design phase is hard, but can reap unexpected rewards later in the design process after the architecture is complete, when a consideration of a variety of solutions implementing the architecture yields a choice that is simpler, more flexible, or more powerful than if the design had been biased toward a solution too early.

Step 4 in the functional architectural process involves making a choice about which interfaces to make programmatic and which to make network-based. The technical criteria that constrain the classification of interfaces as programmatic or network-based often involve performance, access to particular data, or the need for replication and distribution. Performance constraints may dictate whether an interface is programmatic or network-based because protocol exchanges between network devices can require milliseconds or more depending on the network distance between the devices, whereas programmatic interfaces typically require less than a millisecond on modern processors. If a cryptographic algorithm is especially computation-intensive, a network interface may be necessary if the function involving that algorithm is located on a specialized device with dedicated cryptographic hardware. If a particular functional entity requires access to a large amount of data, a programmatic interface may be necessary if the data resides in main memory or in a database on the local disk. If a particular functional entity needs to be replicated at various points in a network, or if the functional entity needs to be distributed to provide reliability and robustness in the face of network failures, a network-based interface may be necessary between the different instances of the functional entity and/or between the functional entity and others.

The difficult part of developing a functional architecture is deciding how to classify the network-based interfaces as open or closed, which is step 5 in the functional architecture process. The technical aspects of system design often do not constrain the decision enough to point to an obvious choice, so non-technical criteria, such as business considerations, often play a major role in deciding which interfaces to open and which to close. If the participants in the design process are willing to honor the technical constraints where they exist, then non-technical criteria are often useful where technical constraints do not exist, since the satisfaction of such non-technical criteria can make vendors and network operators more interested in actually deploying a protocol or system. While it might seem inappropriate to take such considerations into account when doing an engineering design, the reality is that they heavily influence the kinds of wireless network architectures that are standardized and therefore the kinds of protocols that are developed.

Network operators typically like protocol interfaces to be open because they would like the widest possible choice of interoperable hardware, in order to facilitate competition and thus (hopefully) lower prices. Vendors, on the other hand, like closed interfaces because they can be used to lock customers into purchasing complete network systems and not just single boxes. As a result, sometimes the decision whether to make an interface open or closed is governed by a tussle between operators and vendors in the design and standardization process. If the interface is between new network entities and older ones, and the interface to the older ones is either standardized or proprietary, then the decision is clear – the protocol on the older interface must be matched on the new network device. The wireless interface between network equipment such as access points and base stations and end-user equipment such as handsets and interface cards is usually open, since even though there are vendors that manufacture both end-user equipment and network equipment, the vendors want their end-user equipment to interoperate with other vendors' network equipment.

Closed interfaces are often appropriate where the collection of functions on the interface is thought to be incomplete at the time the initial design is done. Making the interface closed gives vendors an opportunity to experiment with various extensions, which can be standardized later if some prove useful beyond the implementing vendor's application. The danger with closed interfaces is that multiple, incompatible versions of the interface can proliferate, making a later consolidation necessary for achieving interoperability. This situation can hinder initial deployment.

2.3   Example functional architecture for a simple wireless system

In this section, we develop a functional architecture for a simple wireless system: the wireless key fob, offered with many late-model automobiles. The key fob allows a driver to open the doors remotely while still walking to the car. On the face of it, using an architectural approach to design a system which is so simple and really well known from everyday use might seem a little silly, but the simplicity and familiarity has advantages. Simplicity means that we can discuss the architectural approach in a couple of pages and not get bogged down in excessive detail. The familiarity means that goals of the system and the functionality are fairly clear.

2.3.1   System goals

For such a familiar system, the system goals should be well known. The system should:

- allow the user to remotely lock the car
- allow the user to remotely unlock the driver's side door
- allow the user to remotely unlock all doors
- allow the user to remotely activate the horn and headlights to help the user find the car
- cause the horn and headlight display to cease on activation of any other control or opening the doors with the physical key, if the horn and headlight function has been activated.

These goals are very high level, general, and also qualitative. Perhaps after review some quantitative constraints seem desirable; for example, the maximum amount of time between when the user activates a function and when the car responds, or a maximum duration for the "panic button" functionality in the fourth goal to avoid annoying neighbors by long periods of unconstrained honking. But the goals in the above list should be sufficient for demonstrating the next step, determining the system functions.

2.3.2   Required system functions

Based on the system goals, we can now draw up a list of system functions. Here is the list (the functions are numbered for further reference below):

1. Activate signaling upon button 1 press to lock car.
2. Receive signal to lock car.
3. Send lock command on car's command bus to all doors.
4. Activate signaling upon button 2 press to unlock driver's side door.
5. Receive signaling to unlock driver's side door.
6. Send unlock command on car's command bus to driver's side door.
7. Activate signaling upon button 3 press to unlock all doors.
8. Receive signaling to unlock all doors.
9. Send unlock command on car's command bus to all doors.
10. Activate "panic button" signaling upon panic button press to beep and flash.
11. Receive "panic button" activation signaling.
12. Send beep and flash command on car's command bus.
13. Activate signaling upon any button press to deactivate "panic button" if "panic button" is currently active.
14. Receive "panic button" deactivation signaling.
15. Send beep and flash termination command on car's command bus.

2.3.3   System functional entities

Since we are modeling an existing system, the network entities in the system are clear: the key fob and the car. The functional entities need to be defined in a way that fits into these network entities. The system functions can be grouped into three functional entities. They are:

- an "Activate Signaling" functional entity supporting Functions 1, 4, 7, 10, and 13 from the above list;
- a "Receive Signaling" functional entity supporting Functions 2, 5, 8, 11, and 14 from the above list;
- a "Command Origination" functional entity supporting Functions 3, 6, 9, 12, and 15 from the above list.

The basis for grouping these functions is the similar role the grouped functions play in implementing the goals and the network entity on which they are implemented. Because Functions 1, 4, 7, 10, and 12 are all user-facing functions, they are implemented together on the key fob. Functions 2, 5, 8, 11, and 14 are involved in receiving and interpreting the commands from the user interface, and translating command parameters into internal data structures that can be used to carry out the commands. They are implemented together on the car. Functions 3, 6, 9, 12, and 15 translate internal data structures and command flow from user interface reception to commands on the car's command bus. They are also implemented on the car.

2.3.4   Selection of interface types

Interfaces between the functional entities and their type (network-based or programmatic) are determined by where the entities are implemented:

- N1 – The Activate Signaling entity is implemented on the key fob which also supports the user interface for the system. The Receive Signaling entity is located in the car. The interface between the two entities is a wireless protocol interface.
- P1 – The Command Origination entity has no interface with the Activate Signaling entity, just with the Receive Signaling entity, and it too is implemented in the car. In addition, the Command Origination entity and the Receive Signaling entity communicate so the interface between them should be a programmatic interface.

The next step is deciding which interfaces should be open and which should be closed. Whether or not N1 should be an open interface, for which the specification is standardized and published, depends on the business goals for the product. If there are currently no standards for wireless remote key entry devices, or the car manufacturer wants to keep control over the protocol for business or other reasons, the interface might be kept proprietary. If, on the other hand, the car manufacturer wants to enable an aftermarket business in remote car door opening devices, perhaps for a "convergence" device that supports both car and garage door opening, then the interface should be standardized and published.

2.3.5   Functional architecture

Figure 2.1 shows the functional architecture for the remote car key system. There are two network entities, the remote entry key fob and a (possibly software) module on the car that implements the key fob commands.

2.4   Functional architecture for network security systems

The functional architecture approach to network system design described above is quite general. Applying it specifically to security systems requires a few specializations of the approach. For security systems, the security goals are typically derived from the threat analysis. Since the security system is intended to counter threats uncovered during the threat analysis, each specific threat should generate a goal involved in countering the threat. There may be additional goals for the security system that are unrelated to threats but necessary for other reasons. For example, if the security system interfaces with other network subsystems and protocols, goals constraining the design to accommodate the interaction between the security system and other components are necessary. In addition, the functional architecture of the network subsystem for which the security system is being designed may constrain the security architecture in other ways, since the security architecture must be designed to address threats to the network architecture.

A good threat analysis should be specific enough to constrain the definition of security system functions, but not so specific as to limit the applicability of the functions too sharply. For example, threats to confidentiality may arise from a variety of sources – passive eavesdroppers, active attackers, etc. Listing each of these as a separate threat might

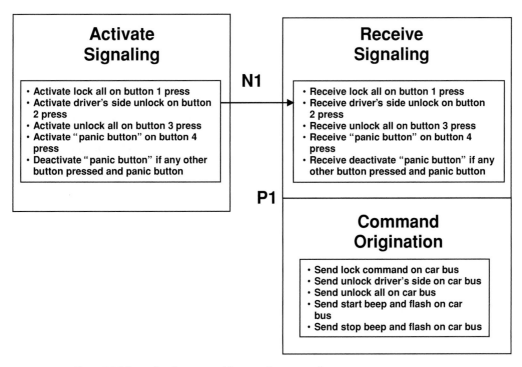

Figure 2.1 Network reference architecture for remote key system

lead to separate functions for confidentiality to counter each threat. Unless the nature of the threat to confidentiality is fundamentally different, all threats to confidentiality should be grouped under the same heading. Fundamental differences between threats within the same class of threat are usually evident when there are basic differences in the security prerequisites, for example, if the pre-provisioned cryptomaterial (keys, certificates, passwords, etc.) must be different or if different algorithms must be used. Sometimes, these differences are generated by backward compatibility requirements necessary to accommodate pre-existing security system components.

After the threats have been identified, the following steps result in a security architecture:

- Select a security service from those listed in Chapter 1 to mitigate each threat.
- If any supporting systems are needed, select supporting systems from those listed in Chapter 1. If supporting systems are available from existing security system components, then use them.
- Develop functions around the security services and the supporting systems.
- Define functional entities and interfaces.

For example, suppose there is a threat to a communication session between two parties involving third-party eavesdropping. A function included to counter that threat involves data confidentiality protection. A supporting system for key distribution may be required if the existing security systems do not have key distribution support. After the functions have been defined, functional entities and interfaces between them and between programmatic components are defined to complete the functional architecture.

Often, the choice of functional entities and interfaces is dictated by the overall network system architecture or by existing network entities. For example, suppose there is a requirement for confidentiality protection between two communicating network entities in the network system architecture. Rather than define a new network entity, the security architecture simply adds additional functions to the existing communicating entities. In the process of designing the security architecture, certain threats may be identified that require modifications to the network architecture. The process is not linear, iteration is sometimes necessary to ensure that the network architecture supports the security architecture.

While it is generally best not to specify cryptographic algorithms at the functional architecture stage if at all possible, other requirements independent from the security requirements usually dictate what particular type of cryptographic algorithm to use. In Chapter 3, we discuss specific types of cryptographic algorithms with examples that are widely used in wireless Internet protocols. The network architecture for the subsystem under design may dictate what type of cryptographic algorithm is best. For example, a network protocol that involves a wireless terminal with no prior business or user relationship interacting with the access network may require public key cryptography rather than shared key cryptography because the two sides do not have a preshared secret. These considerations require that the requirements of the cryptographic algorithms need to be incorporated into the security system functional architecture.

2.5 Example functional architecture for a wireless security system

As is typical, we developed the network architecture first for the key fob. In this section, we apply the process described above to develop a security architecture for the key fob. The security architecture does not add any additional network entities, but it does require some additional functions and functional entities on both the car and the key fob itself. Also, we assume there are no existing security subsystems, though, in practice, the car may support some security systems to control access to critical engine components over the car's bus. The sections below step through the security architecture development process.

2.5.1 Identify the threats

The communication between the key fob and the car is the primary target of interest for an attacker, and is the most dangerous because the attacker can access it while some distance away from the owner or the car. While it is possible for an attacker to target the car or the key fob, typically the key fob will be in the owner's possession and the car will be protected by locking. Attackers can obtain access to either only if they access the physical object. Below, we examine each threat category from Chapter 1 for possible attacks on the fob to car communication.

- Replay attack: The attacker replays captured traffic in order to cause the car to unlock, thereby gaining entry to the car. This is clearly a threat, though in order to actually capture the traffic, the attacker needs to be within a short (and thereby possibly visible) distance of the car owner, because the range of the key fob is limited.
- Eavesdropping: An eavesdropper on the fob to car communication is able to obtain information about when the car owner locks or unlocks the car. Because the radius of communication is typically short, the attacker has to be within a short distance of the car owner in order to obtain this information. Since the attacker could also determine the owner's actions by simply watching what the owner is doing, an eavesdropping attack alone on the fob to car communication is probably not a serious problem.
- Spoofing: If the attacker can spoof traffic from the key fob, unauthorized access to the car can be obtained. A spoofing attack is particularly serious if it can be launched by an attacker that has never had access to any communications between the key fob and the car, because this would allow the attacker to perpetrate the attack without ever being in a potentially vulnerable position near the key fob, where its presence could be detected. This is clearly the most serious threat to the key fob system.
- Man-in-the-middle attacks: A man-in-the-middle attack occurs if the attacker can position itself between the key fob and the car in the communication. Depending on what occurs with the traffic, the man in the middle attack could be more or less of a threat. The attacker could use its position to launch a denial-of-service attack, effectively preventing the owner from opening the car. The attacker could analyze the intercepted traffic in order to try to crack security, or could replay the traffic in order to gain access to the car. Of these, the replay attack is probably the most important, as discussed above. Denial-of-service attacks are discussed in the next paragraph. While traffic analysis is clearly a problem, in this case, the attacker cannot derive much useful information from the traffic other than what is already known, unless the cryptographic algorithms used to secure the traffic are insufficiently robust.
- Denial-of-service attack: The attacker can launch a denial-of-service attack by capturing and replaying valid traffic, as a man in the middle. If the protocol or implementation has any bugs, this could lead to crashing the car's operating system or causing the key fob to lock up. These kinds of problems can be addressed by carefully designing the protocol implementation to avoid buffer overflows and other obvious sources of security problems, and then stress-testing the implementation to help identify bugs. Of course, there are easier ways for the attacker to launch a denial-of-service attack, for example, turning on a radio noise generator on the frequency of the key fob.

2.5.2 Select security services to mitigate the threats

The threat analysis reveals that the two most serious threats involve spoofing. The most serious threat is an attack in which the attacker can fabricate a spoofing key fob without having access to any network traffic and thereby open the locked car. The next most serious threat is when the attacker can insert themselves as a man in the middle, then gather traffic that will allow opening the car at a later date either by replaying the traffic or by synthesizing a new message. In contrast, threats that may require confidentiality seem to be of lesser concern. Due to the short range of the key fob, the attacker must be near the key fob owner, so the attacker could derive exactly the same information by simply watching the owner's actions (getting into the car after having opened it, etc.). DoS attacks on the protocol are similarly of lesser concern, since the attacker could mount an attack more effectively by simply starting up a radio noise generator. Nevertheless, good protocol engineering and testing is necessary to ensure that obvious DoS attack bugs – like buffer overflows – don't creep into the implementation, thereby allowing an attacker to crash the car's operating system.

The spoofing attacks suggest that the key fob architecture requires three security services:

- identity authentication to identify that the key fob is, in fact, authorized to act as a key for the car;
- anti-replay protection to prevent the attacker from replaying legitimate messages to open the door;
- data origin authentication to ensure that the messages originate from the authorized key fob.

In practice, since the key fob protocol is a one-shot protocol (one message is sent and received per action), the protocol does not involve session establishment, allowing the identity management and data origin authentication to be combined. Each message is a separate session and therefore no session-specific key provisioning is required.

2.5.3   Select necessary supporting systems

The requirement for terminal and data origin authentication generates an additional requirement for credential provisioning and key management between the car and the key fob at some point prior to the key fob's use as a key. This procedure is outside the basic key fob network architecture described above. It could be done by pre-provisioning credentials on the key fob at the factory, or it could be done as part of an initial "introduction" between the key fob and the car, in which the user or possibly the car dealer performs some kind of activation procedure on the key fob and car to program both with the proper credentials. Either case requires functions in the security architecture to provision the credentials.

2.5.4   Develop functions around services and supporting systems

Applying the solution approaches to the architecture in Figure 2.1 leads to the following set of functions on the key fob:

- Key Fob Credential Configuration – this function runs prior to the key fob being used as a key and configures the key fob with credentials whereby the key fob can authenticate itself with the car.

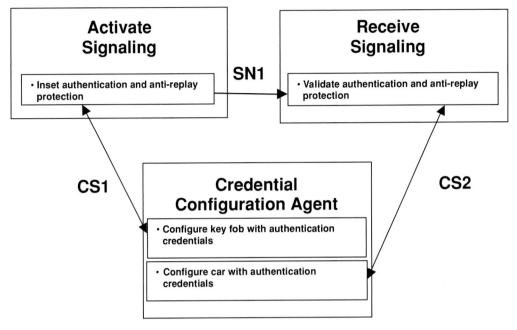

Figure 2.2 Security architecture for the key fob example

- Signaling Authentication and Anti-Replay Protection – this function insets authentication and anti-replay protection on the messages from the key fob to the car.

The matching functions on the car are:

- Car Credential Configuration – this function runs prior to the key fob being used as a key and configures the car with a set of matching credentials allowing the car to authenticate the key fob.
- Signaling Authentication and Anti-Replay Verification – this function verifies the authentication and anti-replay protection on signaling from the key fob, rejecting any signaling that does not pass verification.

In addition, a function is necessary for co-ordinating the configuration of credentials on both the car and the key fob:

- Credential Configuration Co-ordination – this function is responsible for co-ordinating the pre-use configuration of matching authentication credentials on the key fob and car.

2.5.5   Define network entities and interfaces

Figure 2.2 contains a functional architecture diagram. The security architecture adds one additional functional entity to the architecture:

- Credential Configuration Agent – this is responsible for configuring matching credentials on the key fob and car prior to use of the key fob as a key.

The interfaces for the security architecture are:

- SN1 – data origin authentication and anti-replay protection on the N1 interface.
- CS1 – credential configuration and key management between the key fob and the Credential Configuration Agent.
- CS2 – credential configuration and key management between the car and the Credential Configuration Agent.

The Credential Configuration Agent contains the Credential Configuration Co-Ordination function. As mentioned previously, the Credential Configuration Agent may require human intervention, and may not be automatable, depending on the credential configuration algorithm.

2.6   Summary

In this chapter, we discussed a functional architecture approach to wireless network security system design. The functional architecture approach begins with a list of goals that the system needs to satisfy. For a security system, the security goals are derived directly from the threat analysis. Additional goals may be necessary to accommodate other aspects of the system such as backward compatibility with existing security subsystems. The goals then lead to the definition of a collection of system functions that implement the goals. The system functions are then grouped into functional entities based on criteria such as what external systems the functions must interact with, what functions must interact closely together, what functions play similar roles in the system, etc. Interfaces between the functional entities – where functions within the entities interact or where functions within the system interact with external systems – should then be clear. The interfaces are classified as programmatic, open network, or closed network depending on whether the interacting functional entities are distributed or local, and whether the protocols on the interfaces are to be open and standardized or closed and proprietary. When the network interfaces are clear, the functions communicating across those interfaces and their parameters define the semantics of the protocol. The final step in developing the architecture is to group the functional entities into functional entities, which are mapped into existing or new network entitles. The final step essentially determines what network products will be available to build an actual, functioning wireless security network.

3 Cryptographic algorithms and security primitives

Wireless security is built on a collection of cryptographic algorithms and security primitives providing the algorithmic underpinnings for the security services and supporting systems discussed in Chapter 1. The same algorithms that are used for wireless security are also used for Internet security in general. The differences primarily stem from how the algorithms are used in wireless and mobile Internet systems, which is the topic of future chapters. In this chapter, we review security algorithms and primitives that are common to both wireless security systems and Internet security systems in general.

Much of the material in this chapter is available from other sources in more detail than presented here. The material here is intended to present an overview of the cryptographic algorithms and security primitives commonly found in wireless Internet security systems. Before selecting an algorithm for design work, however, a more complete reference should be consulted. It is particularly important that the vulnerabilities of the algorithms are well understood. Uncompensated weaknesses or algorithms that are used in an inappropriate manner may result in opportunities for attack. Detailed information about the cryptographic algorithms and security primitives discussed in this chapter can be found in the books (Menezes, Oorschot, & Vanstone, 1997) and (Kaufman, Perlman, & Speciner, 2002). In addition, Wikipedia is an excellent reference on different cryptographic algorithms, for example (Wikipedia, 2008b) on the RSA public key algorithm. Wikipedia pages can be found simply by searching online in Wikipedia using the algorithm name as the key.

The five different classes of algorithms examined in this chapter are:

- anti-replay protection algorithms
- message digests and cryptographic hash functions
- shared key cryptographic algorithms
- public key cryptographic algorithms
- secure key provisioning.

The mapping between these algorithms and the security services and supporting systems described in Chapter 1 is fairly straightforward:

- Replay protection uses anti-replay protection algorithms.
- Data origin authentication and key management use message digests and cryptographic hash functions.

- Data origin authentication, confidentiality protection, and certain types of key management use shared key cryptographic algorithms and public key cryptographic algorithms.
- Key management uses secure key provisioning primitives.

We discuss each of the classes in the following sections.

3.1   Replay protection algorithms

The simplest class of security algorithms is anti-replay protection. Anti-replay protection ensures that an attacker cannot intercept a message sent to a legitimate recipient and replay it at a later time. Anti-replay protection algorithms fall into three classes:

- sequence number
- nonce
- time stamp

Sequence numbers are used when a network session continues over a longer period and involves an exchange of many messages. Starting with a low number, the sender puts an increasing sequence number on each message. The receiver keeps track of the sequence numbers and ignores any number that is lower than the currently active one. Since sequence numbers have a limited number of bits, special measures are necessary when the sequence number rolls over. Either the two sides need to reinitialize the session or some special signaling is required to indicate that the new message is legitimate even though the sequence number is lower than the previous number.

A *nonce* is a randomly generated number that is attached to a request message and to the corresponding reply. Nonces are used in request/reply protocols where a single request from the sender is matched by a single reply from the responder. The sender randomly generates a nonce and attaches it to the request. The sender keeps track of the nonce while awaiting the reply. The responder includes the nonce in the reply. The sender knows the reply matches the request sent by matching the stored nonce against the nonce in the reply. While an attacker can record a message sent by the responder and replay it to the requester, the requester will ignore old messages because the nonce does not match an outstanding request.

A *time stamp* is another way to protect a request/response protocol from replay attacks. The sender or responder determines the current time, obtained from the Network Time Protocol (NTP) (described in RFC 1305 (RFC 1305, 1992)) or from some other well-known source. Both sender and responder include the current time in their messages. A message recipient drops the message if the time stamp is outside a particular window around the current time. The window is necessary because clocks on individual nodes in a distributed system cannot be synchronized exactly. Time stamps are a bit less secure than nonces and sequence numbers since both sides are vulnerable to a replay attack for messages that are caught and replayed within the window. Time stamps are typically more useful for protocols in which there is little delay on messages sent between the two sides, (i.e. the responder is topologically near the requester in the network) so the window can be kept small. In addition, time stamps can fail if the clocks on the two nodes involved are badly out of sync. An advantage of time stamps is that protocols in which the sender periodically multicasts or broadcasts a message to many other nodes can use time stamps.

3.2 Message digests and cryptographic hash functions

Data origin authentication requires the sender of a message to construct a cryptographic proof of origin, based on the contents of the message. The proof allows the receiver to verify, with a high degree of confidence, that the message actually did originate with the claimed sender, and that the message was not modified in transit. A message requiring authentication might be many bytes long. If the construction of the proof requires as many or more bytes as the message itself, the resulting communication bandwidth between the two parties is effectively cut in half. While such a reduction in communication bandwidth is undesirable in any case, it is especially problematic for wireless communication, where conserving scarce and expensive bandwidth over the wireless link is particularly important.

Cryptographic hash functions and message digests were developed as a way to reduce the amount of data necessary to construct cryptographic proofs of data origin. A *message digest* is a small amount of data (typically 128 bits or 16 bytes) that summarizes a message in a way that is impossible to forge and difficult to duplicate except from the bytes of the original message. A *cryptographic hash function* is a noninvertible function that maps the bytes in a message to a unique message digest. When a shared key is an argument to the function in addition to the bytes of the message, a cryptographic hash function is often called a *keyed hash*. A message digest formed using a keyed hash from both the message and a secret key shared between two parties is called a *message authentication code* (MAC).[1] Shared key MACs are used widely in Internet security protocols for data origin authentication. Message digests can be used with public key cryptography too, as we will see later in the chapter.

3.2.1 Important properties of cryptographic hash functions

In many ways, a cryptographic hash function is exactly like the hash functions used for content-indexed storage in hash tables. A *hash function* maps an input key to mostly a unique output hash value. When two inputs map to the same output, a collision occurs. Collisions are typically not a problem in a hash table as long as they are not frequent. A hash table stores colliding items and their keys in a hash chain. If a key collision

---

[1] Note that we use the abbreviation "MAC" to indicate "message authentication code" throughout the book, as is typical in the network security literature. The radio link layer protocol literature uses MAC for "Media Access Layer" and "MIC," for "Message Integrity Check," to indicate a cryptographic hash used in data origin authentication. We use the term "link layer" for all protocols below the Internet Protocol (IP) layer, including the media access layer.

occurs on lookup, the corresponding item is looked up by comparing the colliding key sequentially to keys in the chain until a match occurs. However, when hash functions are used for message digests, the probability of a collision must be kept extremely low, since the digest is meant to uniquely identify the message to the recipient. If collisions occur frequently, they lead to misidentification of the message and can be exploited by attackers to substitute one message for another with a matching hash value.

Another important property for cryptographic hash functions is randomness in the output. One measure of randomness is that, given a large enough sample of message digests, the probability of a particular bit being on or off should be about 0.5. Another is that a particular message digest should have a probability near 1 of about half the bits being on, on average. Also, given two inputs with one or a few bits difference, the outputs of the hash function should be uncorrelated. Of course, because a particular input to a hash function must always generate the same output (otherwise the hash function would be useless for message identification), the output cannot be completely random, but it should be close enough that a test for a lack of randomness would require many samples.

Randomness helps ensure that the hash function is noninvertible. Noninvertiblility requires that given an output from the hash function, it is not possible to find the input, nor is it possible to find another, colliding input that could be used for an attack. If the output looks like a random number, then it provides no clue to an attacker about how to determine what the input is or what a colliding input might be. If there is some correlation between the bit patterns in the input and output, or between outputs given different inputs, an attacker can use the correlation to advantage. Cryptographers use the term "preimage resistance" to describe noninvertability when applied to a particular input, and "second preimage resistance" when applied to inputs that collide.

Even with perfectly random output, however, a message digest is subject to guessing attacks. The attacker uses brute force to generate messages, compute the message digest, and compare the message digest with the digest from a captured message. If such an attack can be accomplished while the attacker is on line, perhaps as a man in the middle, the attacker can forge messages to the recipient by substituting a forged message for a legitimate one. Off line guessing attacks are less problematic for message authentication, since they cannot be used to disrupt ongoing communication. However, the results of a successful attack can be stored and if the same message is sent in the future, the attacker can use the results of a successful attack to forge the message.

Guessing attacks exploit the combinatorial mathematics of drawing from a set with replacement. This kind of attack is commonly called a *birthday attack*, since the same considerations lead to the surprising conclusion that the probability of two people having the same birthday is above 0.5 in a group with as few as 23 people, which is more a lot more probable than one would expect. The attack is also sometimes called a *square root attack*, because the attack is expected to succeed with high probability after drawing on the order of square root of the number of elements from the set. For the above example, the square root of 365, the number of days in the year, is about 20. Birthday attacks do not depend on flaws in a particular algorithm; they are a consequence of combinatorial mathematics, so they must be considered a threat regardless of the algorithm. Also, birthday attacks are not only an issue with message digests and cryptographic hash functions, they must be considered in any security area where cryptographic material contains a limited number of bits.

To reduce the probability of a successful birthday attack, the number of bits in the message digest needs to be sufficiently high. Of course, the difficulty of guessing also depends on the computational resources available to the attacker. As computers have become faster, the number of bits necessary to deter brute-force guessing has become larger. So what constitutes "sufficiently high" has changed as technology has advanced. However, most cryptographers consider 160 bits as sufficient for the foreseeable future. If the message digest has 160 bits, then the amount of effort required for a birthday attack is on the order of 2^{80} operations. In general, if the hash value has n bits, the amount of run time necessary for a birthday attack is on the order of $2^{n/2}$.

Given a hash function, $H()$, these considerations can be summarized mathematically by the following three properties:

- Collision resistance: It is computationally difficult to find two distinct inputs, x and y, with $x \neq y$, such that $H(x) = H(y)$.
- Preimage resistance: Given an output, $z = H(x)$, it is computationally difficult to find the input x that hashes to z.
- Second preimage resistance: Given an input y, it is computationally infeasible to find a second input x such that $H(x) = H(y)$.

Unfortunately, unlike other areas of cryptography, the hash functions in current widespread use were not developed based on fundamental mathematical principles, but rather are the result of intuition and heuristic considerations. As a result, there are typically no mathematical proofs that these properties apply to a particular hash function. Improving the mathematical underpinnings of cryptographic hash functions is currently an active area in cryptography research.

3.2.2 Attack example

As a concrete example of an attack, consider the case where two parties, Alice and Bob, are exchanging messages, but Alice's messages are redirected through a third party, Eve, before being sent to Bob (adapted from Kaufman, Perlman, & Speciner, 2002, Chapter 5). This kind of situation is common for email, where the mail messages are queued on an outgoing mail server before being sent to the actual recipient or to the recipient's mail server. Alice and Bob are the email sender and recipient, and Eve is the incoming mail server which queues Alice's email before she can read it. In an attack scenario, Eve has managed to hack into Alice's mail server and now controls it.

Suppose Bob authenticates Alice's messages by using a shared key MAC that Alice generates using a keyed cryptographic hash. Alice, Bob, and Eve all know the hash function. If the cryptographic hash is too weak – either because the MAC does not contain enough bits or the shared key is not long enough (if a keyed hash is used) – Eve can utilize a simple birthday attack to substitute her own message. When Eve receives Alice's message, rather than simply forwarding it unaltered, she tries different alternatives to Alice's message until she finds a collision on the MAC. Eve then strips off Alice's message and substitutes her own, and forwards the result. Suppose that Eve

Table 3.1 Starting constants for SHA-1 message digest algorithm

Word:	A	B	C	D	E
Value:	0x67432301	0xefcdab89	0x98badcfe	0x10325476	0xc3d2e1f0

is Alice's administrative assistant and is allowed to generate the original message for Alice's approval. In that case, the probability of a successful attack can be increased even further, since Eve can prepare a collection of acceptable messages and attack messages in advance, and tweak the acceptable messages to make a collision easier to calculate.

3.2.3   Example cryptographic hash function: SHA-1

One of the most important cryptographic hash functions in Internet security protocols is SHA-1. SHA-1 is widely used to calculate message digests and for many other applications where a collection of bits needs to be summarized for security or other purposes. No key is necessary to calculate a SHA-1 message digest; only the bits of the message are required. Another cryptographic hash function that is used in many older Internet protocols is MD5. SHA-1 is very similar to MD5 and was developed from MD5 to be more secure. Most new Internet protocols specify SHA-1, and the use of MD5 is now officially discouraged. In this section, we examine the details of how SHA-1 message digests are calculated.

The SHA-1 algorithm takes a message having a maximum length of 2^{64} bits. The algorithm breaks the message into chunks of 512 bits each and produces a 160 bit message digest. Before processing, the algorithm pads the message of length n ($\leq 2^{64}$) to a multiple of 512 bits in the following steps:

1. Add a single bit set to 1 to the original message.
2. If the pad length is greater than 64 bits, add $m - 1$ (≥ 0) bits set to 0, for the smallest m that pads the message out to 64 bits less than a multiple of 512 bits.
3. If the pad length is less than or equal to 64 bits, then no zeros are added.
4. Calculate the value $k = n \bmod 2^{64}$.
5. Fill the remaining 64 bits with the value k.

The algorithm appends the padding with the most significant word preceding the least significant and forms the words themselves with the most significant byte preceding the least significant (i.e. little endean convention).

The algorithm then breaks the padded message into 512-bit blocks, each block being processed separately. The output from each block processing is a 160 bit (20 bytes or five 32-bit words) digest that summarizes all blocks processed up to that point. The output from processing one block serves as the input to processing the next block, and the message digest of the entire message is the five 32-bit words output after processing the final block. The message digest is initialized to the five 32-bit hexadecimal words listed in Table 3.1.

Define *RotateLeft* (X, m) as an operator that rotates the 32-bit word X left by m bits. Before processing the first block, initialize $a_0 = A$, $b_0 = B$, $c_0 = C$, $d_0 = D$, and $e_0 = E$.

Table 3.2 Constants and functions for SHA-1 calculations

	K_i	$f_i()$
$0 \leq i \leq 19$	0x5a827999	$(b_i \wedge c_i) \vee (\sim b_i \wedge d_i)$
$20 \leq i \leq 39$	0x6ed9eab1	$b_i \oplus c_i \oplus d_i$
$40 \leq i \leq 59$	0x8f1bbcdc	$(B_{i-1} \wedge C_{i-1}) \vee (B_{i-1} \wedge D_{i-1}) \vee (C_{i-1} \wedge D_{i-1})$
$60 \leq i \leq 79$	0xca62c1d6	$B_{i-1} \oplus C_{i-1} \oplus D_{i-1}$

The algorithm processes each 512 block in the following way:

1. Create a buffer Q of eighty 32-bit words in the following way:
   a. The first 512 bits (16 words) contain the bits from the message block.
   b. For the rest of the buffer, the nth 32-bit word, starting with $n = 16$, is constructed by XORing the n-2th, n-8th, n-14th, and n-16th words together and rotating left one bit.
2. For $i = 0$ to 79 do:
   a. Let $b_{i+1} = a_i$, $c_{i+1} = RotateLeft (b_i, 30)$, $d_{i+1} = c_i$, $e_{i+1} = d_i$.
   b. Let $a_{i+1} = e_i + RotateLeft (a_i, 5) + Q_i + K_i + f_i (b_i, c_i, d_i)$ where Q_i is the ith element of the buffer Q and K_i and f_i () are defined in Table 3.2. In the table, $\wedge$ indicates bitwise AND, $\vee$ indicates bitwise OR, $\sim$ indicates logical NOT, and $\oplus$ indicates XOR.

This step is repeated for each 512-bit block of the message. After the last 512-bit block is processed, the algorithm outputs the message digest as the concatenation of a_{80} | b_{80} | c_{80} | d_{80} | e_{80}.

A note of caution is appropriate here regarding the use of SHA-1. Despite widespread inclusion in many security protocols, recent cryptanalysis results suggest that the security of SHA-1, like MD5 before it, may not be sufficient in the future. In particular, while the theoretical bound on collision resistance is 2^{80}, a technique has been discovered for finding collisions in 2^{69} operations. So far, there are no known instances of attacks that have been attempted using these results, but there are published examples of likely inputs that could lead to collisions and that are plausible examples of how an attack might take place. Considering the threat, SHA-1 should in any case not be used for digital signatures created with public keys in future protocols but is probably safe for use in keyed hash MACs (such as HMAC described in the next subsection) created with shared keys, at least for a while. The future prospect is that SHA-1 will eventually be replaced with a new message digest function, hopefully with provable security at least in part, since SHA-1 security itself is not proven. The Internet community is currently working on selecting a successor.

3.2.4   Example keyed cryptographic hash function: HMAC

By itself, a message digest does not provide much security. The message digest summarizes the message, but if the message digest is included in a message without any other processing, there is nothing to prevent an attacker from changing the message and calculating a new message digest. Unless the message digest is processed in some way to provide unforgeable cryptographic proof of generation by the actual sender, a message digest provides no secure authentication of data origin.

One way to accomplish such proof is for the sender and receiver of a message to share a secret key and to use that shared key in the calculation with a keyed cryptographic hash function to form a MAC. Both the sender and receiver use the key in their calculations. The receiver verifies the message by matching the cryptographic hash result it calculates directly from the message to the value the sender attached to the message. Since the key does not appear in plaintext in the message, there is no way for an attacker to modify the message without alerting the receiver that a modification has occurred, unless of course the attacker is able somehow to guess the key.

One of the most commonly used MAC calculation algorithms is HMAC. HMAC must be paired with SHA-1 or another message digest algorithm in order to achieve a complete MAC calculation. An important characteristic of HMAC is that it is provably secure, if the underlying message digest algorithm is, for the following two security properties:

- An attacker cannot produce a collision from two differing inputs.
- An attacker cannot compute a digest if the key is not known, even if the attacker can see digests from an arbitrary number of messages.

The second property essentially means that the attacker cannot determine the key from the digest even with an arbitrary number of samples.

The HMAC algorithm is as follows:

1. If the key size n is less than 512 bits, pad the key out to 512 bits using zeros. If n is greater than 512 bits, then digest the key to m bits, where m is the number of output bits in the message digest algorithm (160 for SHA-1) and pad to 512 bits.
2. XOR the padded key with a constant string of bytes having value 0x36.
3. Concatenate the padded key to the front of the message and apply the message digest algorithm.
4. XOR the padded key with a constant string of bytes having value 0x5c.
5. Concatenate the result to the front of the digest obtained in step 3 and apply the message digest algorithm.

The output contains the MAC for the message.

3.3   Shared key encryption

Secret or shared key encryption provides confidentiality protection for message traffic between two parties that have previously arranged to share a secret key. As in the case of a cryptographic hash function, the key is a random, high entropy cryptographic bit pattern that is mixed with the message traffic in an algorithmically predefined fashion to render the traffic into a seemingly random stream of bits. Ideally, an eavesdropper without access to the key obtains no information about the clear text message from the encrypted text, but the receiver, possessing the key, can use the decryption algorithm and the key to extract the clear text message from the encrypted text. Unlike cryptographic hash functions, however, there is no need to summarize the message traffic in order to reduce the size of the transmitted material. The number of bytes in the encrypted traffic is expected to be as large as, and in some cases even larger than, the original clear text.

Shared key encryption algorithms can be loosely divided into two types, block ciphers and stream ciphers. Block ciphers can be either shared key or public key; we discuss shared key block ciphers in this section and public key ciphers in the next. Stream ciphers are used for wireless security at the link layer, while block ciphers are more important for security at the network (IP) layer and above. The subsections below discuss stream ciphers and block ciphers.

3.3.1   Stream ciphers

A *stream cipher* is an encryption algorithm in which the clear text data is encrypted bit by bit or byte by byte rather than in larger chunks. Unlike block ciphers, there are no explicit Internet standards for stream ciphers. Stream ciphers are never used above the link layer for encryption in wireless Internet protocols or systems, except at the application layer for very specific applications. However, stream ciphers are important in some wireless link-layer standards. For example, the notorious Wireless Equivalent Privacy (WEP) algorithm, which was the encryption and authentication algorithm in the original 802.11–1999 wireless standard (802.11, 1999), is based on the RC4 stream cipher. The experience of the 802.11 community with WEP, in which the original design did not accommodate known weaknesses in RC4 resulting in the deployment of a wireless security standard that was relatively easy to attack, should serve as a cautionary tale for anyone working on security protocol standards for the wireless links (see Edney & Arbaugh, 2004 for details).

Stream ciphers are very well characterized theoretically, and are used in cases where buffering is very limited or when incoming traffic is processed on a byte-by-byte basis. When implemented in hardware, stream ciphers tend to be faster than block ciphers and have simpler circuitry. Stream cipher algorithms are usually also proprietary and specialized. Because they are not relevant to wireless Internet security architectures, they are not covered further in this book.

3.3.2   Block ciphers

Shared key block ciphers are fundamental building blocks for wireless Internet security, and are included in many Internet standards. They are used not only for encryption but also for other purposes, such as shared key message authentication codes, shared key digital signatures, and for other kinds of message and network entity authentication. A *block cipher* is a function that maps a fixed-size clear text block into a fixed-size encrypted text block. Ideally, the sizes of both clear text and encrypted text blocks are equal. An increase in the size of the encrypted text can occur if the mapping from clear text to encrypted text is not one-to-one, since additional information must be included in the encrypted message to disambiguate the encrypted text. Most popular shared key block ciphers are one-to-one mappings. Block ciphers often include an *initialization Vector*, which is used to initialize the encryption/decryption algorithm.

The security of a block cipher in actual applications depends on three properties:

- The theoretical and practical security of the block cipher algorithm itself. If the block cipher is based on well-structured mathematics, the probability of compromise can often be bounded using mathematical proofs. Long practical experience based on repeated cryptanalysis attacks over the years also establishes the security bounds in actual use.

- The length and randomness of the key. Typically, a larger key is better because, if the key is drawn randomly from the key space, an attacker will have a harder time brute-force guessing if the key is long. However, if the randomness of the key drawing is suspect, then even a large key may not provide adequate security. Most block cipher algorithms have a recommended minimum key size. Some algorithms have particular sets of keys which are easily extracted from the encrypted text by an attacker or otherwise result in easily compromised encrypted text. These keys must be avoided or the security of the communication cannot be maintained.

- The size of the input clear text block. Smaller block sizes are undesirable because they increase the probability of successful attack, using statistical analysis or if the attacker happens to obtain a sample that maps the clear text to the encrypted text. Extremely large block sizes are computationally inconvenient since they require extensive buffering in the implementation. Since wireless Internet traffic is sent in packets having a maximum transmission unit (MTU) size, the packet MTU size, minus the size of any headers, is the theoretical upper bound on what an Internet wireless security block cipher can support. Most Internet standard block ciphers have block sizes of 64 bits (two 32-bit words or 8 bytes) or 128 bits (four 32-bit words or 16 bytes), considerably below the common packet MTU sizes on the Internet.

3.3.3   Attack characterization

When analyzing the security of a block cipher, the usual assumption is that the attacker has access to all the encrypted data and has full knowledge of the details of the encryption algorithm, including block and key sizes. The only detail of the encryption that remains unknown to the attacker is the secret key, which thereby determines the security of the communication. The effectiveness of an attack depends on exactly what the attack reveals. An attack that only reveals the clear text is only partially effective, since it just compromises the confidentiality of the analyzed encrypted text. Presumably the attacker must continue to perform the same analysis, which is always much more computationally intensive than the actual decryption, in order to crack future messages. An attack that reveals the key, however, is devastating, because it allows the attacker to perform simple decryption on all future encrypted messages until the key is changed.

The strength of attacks against block ciphers depends on the amount of information available to the attacker:

- The most common kind of attack is when the attacker only has access to the encrypted text and any additional information used to initialize the decryption in messages exchanged between the two parties.
- Another kind of attack is where the attacker has access to some mappings between clear text and encrypted text, but exactly which mappings is not under the attacker's control, and the attacker may not be able to see all such mappings.
- The final kind of attack is where the attacker is free to choose the mappings between plain text and encrypted text for cryptanalysis. This could happen if the attacker has managed to compromise a trusted server storing copies of both the plain text and encrypted text.

The first attack in which the attacker has access just to the encrypted attack is relatively easy to mount. The second and third attacks in which the attacker has access to some plain text are harder to mount, because the attacker requires considerably more information than can be obtained by simply eavesdropping on the encrypted conversation. The third attack, in which the attacker can choose the plain text attack and encrypted text for analysis, is clearly the most dangerous, since it allows the attacker to systematically compare and analyze the encryption mapping, and thereby search for the key. On the other hand, since chosen plain text attacks are the most dangerous, security for block ciphers is usually proven or otherwise established for chosen plain text attacks because a block cipher that is secure against a chosen plain text attack is secure against the other two types.

In any case, even choosing a block cipher that is susceptible to a partial analysis is not a particularly good strategy, and any wireless Internet protocol or system should be based on a block cipher and parameters that have known resistance to cryptanalysis under an assumption of maximum available computational resources. In fact, given the continual availability of increasing computational resources under Moore's Law (that the computational resources available to users doubles on the average in 18 months to two years), it often pays to build some Moore's Law protection into the selection of an algorithm in order to ensure that the protocol or system continues to provide secure communication regardless of future developments. The selection of a block cipher algorithm and parameters should look for known resistance under the assumption that an attacker will have access to increasing computational resources in the future. In addition, past experience has shown that clever cryptanalysis often reveals new and successful attacks against block ciphers that were previously thought to be secure. These considerations suggest that the parameters and even the block cipher for confidential communication should be negotiable on a case-by-case basis, so that the two parties can upgrade the security of their communications depending on the current best estimate of computational resources available to an attacker and knowledge of successful attacks.

3.3.4 Example shared key block cipher: Advanced Encryption Standard (AES)

Many kinds of shared key block ciphers have been proposed in the literature, but there are really only three that are of importance in wireless Internet protocols: DES, Triple DES, and AES. The Digital Encryption Standard (DES) was published by the National Institute of Standards and Technology (NIST) in 1977 for use in commercial and unclassified government applications. DES specifies what looks like a 64-bit key, of which only 56 bits actually contribute to the security. The other 8 bits contribute an odd parity bit in each byte of the key. There are 16 DES keys that are weak and not recommended for use.

When DES was originally proposed and for many years thereafter, DES was thought to be secure because it was thought highly unlikely that an attacker would have sufficient computational resources to crack DES. A detailed analysis at the time estimated that it would cost $20 million to build a machine to crack DES, and require 12 hours, if a clear text/encrypted text pair was available. By 1998, however, a small nonprofit group was able to build a DES-cracking machine for about $250,000, of which $100,000 was nonrecurring design cost. The machine was able to find a DES key in 4.5 days.

The cryptography community was well aware of the deterioration in DES security prior to 1998, however, and the initial response was to standardize a variation of DES that performed a DES encryption with one 56-bit key, followed by a decryption using another 56-bit key, followed by another encryption using the same 56-bit key as in the first encryption. The overall decryption performs the inverse transformation. The result is called Triple DES and it increases the effective number of bits in the key to 112. An encrypt-decrypt-encrypt sequence with two keys was selected, rather than the more obvious two key encrypt-encrypt sequence or adding more keys and more operations, due to some subtle attacks and possible negative impacts on particular applications. Since DES itself is not particularly efficient to implement in software (though hardware implementations are efficient), Triple DES is even less so. The result was not particularly satisfying.

In 1997, NIST announced an open competition to design a new encryption standard. The competition generated a variety of excellent proposals from leading cryptographers around the world. The proposals were widely discussed, and in 2001, NIST standardized the Rijndael algorithm, developed by Dr. Joan Daemen and Dr. Vincent Rijmen of Belgium, as the Advanced Encryption Standard (AES). The original Rijndael algorithm allows the block size and key size to be chosen independently from 128, 160, 192, 224, and 256 bits; but AES fixes the block size at 128 bits and allows a choice of key size from 128, 192, and 256 bits. An important advantage that AES has over DES is that it is based on some very elegant mathematical results from number theory, so cryptographers have a better idea about the theoretical bounds on AES security. In addition, AES has no known weak keys. AES is used in a wide variety of wireless link and Internet standards, including the 802.11–2007 wireless LAN standard (802.11, 2007).

3.3.5   AES algorithm outline

AES defines two parameters that must be set in a particular implementation. One parameter is the key size. As mentioned above, in AES, the block size is fixed at 128 bits, or four 32-bit words. The key size, in units of 32-bit words, can be 4, 6, or 8. For most wireless applications, 4 word (128-bit) keys should be sufficient. The other parameter is the number of mixing rounds performed. The rounds are used to mix the bits from the clear text data with themselves and with the key, to randomize the clear text. The number of rounds is a function of the block size and key size. More rounds are required for larger-sized keys in order to increase the difficulty of an attack. If the key size is 128 bits, the algorithm has 10 rounds, if the key size is 192 bits, the algorithm has 12 rounds, and if the key size is 256 bits, the algorithm has 14 rounds.

AES maintains an array of bytes as its state. The array has four 32-bit (4 byte) columns. The state array is initialized by filling the array column by column from the 16-byte input block, sequentially rotating the 4-byte words from the input right to left, to generate the columns in the state array. Each round consists of four separate operations on the state array in the following order:

1. A round key taken from the round key array is combined with the bits in the input using bitwise XOR.
2. Bytes in the input are replaced one-to-one with bytes from a table called the S-box.
3. A rotation of rows in the state array that shifts bytes to the right in the last three rows of the state. Bytes that would fall off the right side are moved to the left. Starting with the second row, the bytes are shifted by 1, 2, and 3, for the second, third and fourth row, respectively.
4. A mixing operation that mixes bytes from the rows. This operation corresponds mathematically to polynomial multiplication modulo a fixed polynomial, but it can be implemented using simple table lookup in an extended table, different from the S-box. The mixing operation is not applied to the last round; instead, another iteration of (1) is performed.

The round key array is constructed from the shared key by expansion. The expansion generates an array with 4 byte columns containing a total of 44, 52, or 60 columns depending on whether there are 10, 12, or 14 rounds. The expansion starts by first rotating the key, word by word, into the key array columns until all k 32-bit words in the key are used ($k = 4$, 6, or 8), filling the first k columns. Do one of the following to generate additional columns:

- If the column is a multiple of k, first do the following on the previous word:
  1. Rotate the previous word one byte to the right.
  2. Replace bytes in the word with bytes from the S-box of Step 1 in the mixing algorithm.
  3. XOR the result with an array of round constant bytes.
  4. Perform the S-box substitution on the word.
- If $k = 8$ and this is the fourth column in the set of k columns, first perform the S-box substitution on the previous word.
- Use the previous word without any transformation.

Finally, XOR the result with the word k columns earlier.

The round key generation algorithm terminates when 44, 52, or 60 columns have been generated, depending on the number of rounds, which might be in the middle of a set of k columns.

The decryption can be performed by a straightforward inverse of the encryption algorithm, with different S-box and mixing lookup tables. There is also a more efficient form of decryption that takes advantage of the commutative property of some operations. This allows certain operations to be rearranged to increase the performance of the decryption.

3.4   Public key algorithms

Public key or asymmetric cryptographic algorithms are based on a different set of operating procedures and mathematics than shared key algorithms. Rather than having a shared key, the participants in the security service transaction have a pair of keys: a public key, which they distribute widely, and a matching private, secret key, which they keep to themselves. Each participant in the transaction has a separate key pair. The two keys are used in different ways for data origin authentication and confidentiality protection. The sender of a message initiates data origin authentication by forming a message digest and calculating a *digital signature* from the message digest using the private key. A digital signature serves the same function in asymmetric key systems as the MAC in symmetric key systems. The receiver verifies the digital signature by operating on the digital signature using the sender's public key. With encryption, the sender uses the receiver's public key to encrypt the message. Only the receiver's private key can decrypt the message so the message is secure in transit.

The advantage of public key algorithms is that distribution of the public key is considerably easier than distribution of a shared, secret key, since extraordinary measures are not required to prevent outsiders from learning the key. The only requirement is that the recipient of a public key for a particular network entity must be able to verify the authenticity of the key; in other words, that the identity of the network entity which is reputed to possess the matching private key does, in fact, match the network entity that possesses it. A disadvantage of public key algorithms is that they tend to be considerably slower than shared key algorithms, and to require more complex arithmetic processing. A common practice is to use a public key to encrypt and distribute a shared key, which is a relatively small data item. The shared key is then used for bulk encryption or authenticator generation, thereby leveraging the advantages of both types of cryptographic algorithms where they are strongest.

Public key algorithms are based on what is sometimes called a mathematically "hard" or intractable problem from number theory. The difficulty of the problem is measured by how computationally feasible it is to solve the problem in polynomial time. If a considerable number of instances of the problem can be solved in polynomial time, then the problem is not considered sufficiently intractable to be a solid foundation for a secure algorithm. In some cases, the intractability of the problem is provable; in others, it is only inferred. Most of the commonly used public key algorithms are based on one of the following hard problems:

- The integer factorization problem: Given a positive integer, p, find a prime factorization, $p = q_1^{n_1} q_2^{n_2} \ldots q_k^{n_k}$, with the q_i pairwise distinct and $n_i \geq 1$. The difficulty of prime factorization is the basis of the RSA algorithm, one of the most widely used public key algorithms.

- The discrete logarithm problem: Let p be a prime integer and q be a prime integer divisor of $p - 1$. Let G be a set of integers that is a subset of $\{n | 1 \leq n \leq p - 1\}$, $n^q = 1 \pmod{p}$, and such that q is the least positive such integer in G. Then the discrete logarithm problem is the following. Given $p, q, a, b \in G$, find the unique integer x, $0 \leq x \leq q - 1$, such that $a^x = b \pmod{p}$. The difficulty of finding the discrete logarithm is the basis of Diffie–Hellman key exchange, El Gamal encryption, and other algorithms.

In addition to algorithms for the security services, public key algorithms also have specific procedures for key generation. Key generation is typically the most time-consuming operation but, since it is only done rarely, the expense is amortized over lots of security service operations. All public key generation algorithms require the ability to pick large prime integers that have good randomness properties. The size of the integers is larger than the typical 32-bit or 64-word sizes of most computer memory architectures, so the integers will not fit into a single word. Larger-sized integers with good randomness properties make key guessing more difficult, but the lower limit for good security depends on the particular public key algorithm. The exact performance of a public key algorithm and which security service operation is more time consuming depends on the key size and the size of other parameters in the algorithm, as well as such implementation and such operational factors as the performance of the software implementation, processor speed, etc.

Public key algorithms are used for data origin authentication, confidentiality protection, and secure key exchange. Data origin authentication and confidentiality protection are covered in the next two subsections, key exchange is covered later in the chapter.

3.4.1 Data origin authentication

The sender of a message authenticates the message by calculating a digital signature on the message with its private key while the receiver verifies the message by operating on the digital signature using the sender's public key. A more detailed outline of how a digital signature is calculated and verified is the following:

1. A cryptographic hash algorithm, such as SHA-1, is used to calculate a message digest. This step is common with calculation of shared key message authentication codes.
2. The message digest is encrypted using the originating party's private key, forming the digital signature.
3. The digital signature is appended to the message and sent to the other party.
4. The receiving party, which has the sending party's public key, uses the public key to decrypt the message digest.
5. The decrypted message digest is compared with a message digest calculated directly from the message. If the two match, data origin authenticity has been verified.

Since only the sender has access to the private key and therefore could have signed the message, the receiving party can have full confidence that the data originated with that party and was not changed in transit.

The algorithmic details of how the signature is calculated are common with encryption for confidentiality, except in that case, the originating party uses the receiving party's public key to encrypt, since the originating party is not the holder of the private key.

Because they are more computationally intensive than shared key algorithms, public key algorithms are not used for authentication of long message exchanges. This is especially true for wireless devices which tend to have less computational resources due to power and size constraints. The typical pattern is to use public key authentication for the initial contact between two parties, during which a public key algorithm is also used for exchanging shared keys. The shared keys are then used for further authentication and encryption on messages.

3.4.2   Confidentiality protection

Public key algorithms can be combined with a block cipher algorithm to provide confidentiality protection. Instead of using a shared key algorithm for encrypting the blocks, a public key algorithm is used. The sender encrypts the data with the receiver's public key, and the receiver uses its private key to decrypt. As mentioned above, public key algorithms are typically not used for bulk data encryption, since they are substantially more computationally intensive than shared key algorithms. A common use for public key encryption is to encrypt a shared key, which is then used for bulk encryption. Key provisioning using public key algorithms is discussed later in the chapter.

3.4.3   Example public key algorithm: RSA

One of the most commonly used public key algorithms is RSA. RSA was invented by Ron Rivest, Adi Shamir and Leonard Adleman (hence the name) in 1978. RSA is based on the difficulty of integer factorization. For many years, RSA was under patent protection but the patent expired in 1997 and since then, RSA has become the public key algorithm of choice for most Internet public key security applications. In RSA, encryption is more expensive than decryption, but not excessively so, making the efficiency of RSA good compared with other public key algorithms. The three public key operations are implemented by RSA as follows:

Key Generation: Choose two large primes, p and q. For good security, the size should be 1024 bits minimum. Multiply the primes together to form n. Choose a positive integer greater than 1, a, that is relatively prime to $\phi = (p - 1)(q - 1)$. Compute the unique positive integer $1 < d < \phi$, such that $ad = 1(\mathrm{mod}\phi)$. The public key is (n, a), and the private key is (n, d). If the public key is (n, a), the maximum message value, m, that can be encrypted is $n - 1$.

Encryption: Given a message, m, if the intended application is confidentiality protection, the public key is used to encrypt the message: $c = m^a(\mathrm{mod}\ n)$. If the intended application is data origin authentication, the private key, d, is used instead of a. The encrypted text c is sent to the receiving party.

Decryption: Given an encrypted message, c, decryption follows by performing the inverse operation. If the intended application is confidentiality protection, the private key is used to extract the message: $m = c^d(\text{mod } n)$. If the application is data origin authentication, the public key, a, is used instead of d.

The algorithm description is deceptively straightforward. In practice, there are a lot of tricks to implementing RSA that are not apparent from the algorithmic description. For example, the algorithm involves calculating large powers of integers. If this is done in the simplest way possible, the efficiency of the algorithm is severely compromised. In addition, there are a few fairly esoteric attacks that depend on picking specific values for the parameters, or pre-processing the message blocks improperly. In commercial and OpenSource implementations of RSA, such as OpenSSL, the implementations use efficient big number arithmetic for calculations and measures are taken to mitigate the threats. Nevertheless, it is important when using RSA in an application to understand the details of the threats to avoid inadvertently choosing some combination of parameters, however improbable, that triggers one of the weaknesses.

Since the complexity of the underlying "hard problem" for RSA – factoring a large integer – is not proven, RSA is potentially subject to Moore's law risk, particularly from specially designed parallel computers. As the amount of computational resources available becomes larger, the ability to brute-force attack RSA at smaller key sizes becomes easier. A key size of 1024 bits or larger is recommended for this reason. Also, it is known that the development of practical, deployable quantum computers, which perform massively parallel computations at very little cost, would completely compromise RSA at any key size. There are also other public key algorithms that can be used should RSA one day prove insufficient; some of these have proofs of computational complexity, though implementation and business concerns limit widespread deployment.

3.5   Key provisioning

An important part of providing usable security services is the provisioning of the participants with the state supporting successful cryptographic operations using the chosen cryptographic algorithm. Because the primary state necessary for performing jointly intelligible cryptographic operations is a key, this operation is often called key establishment or key provisioning. Key provisioning, along with managing the key over time, is a function of the supporting systems discussed in Chapter 1.

In shared key algorithms, key provisioning requires that the parties end up with a mutually shared key, and that the provisioning process allow no possibility for any other party to learn the key. In public key algorithms, each party generates their public/private key pair autonomously, and arranges for the public key to be available for the other parties. In both cases, the parties exchanging keys require mutual authentication prior to the key provisioning; otherwise, it is possible for one party to end up believing that it has established a key with a valid participant in the protocol, but the other party is actually a spoofing attacker intending to steal or disrupt traffic. Secure key exchange therefore requires a method of securely establishing the identity of the other parties in the exchange, and, in the case of shared key protocols, a method of actually securely generating and/or sending the shared key to the other parties. To simplify further discussion, we assume that the key exchange operation involves only two parties.

Mutually establishing the authenticated identity of the participants is the first step in the key exchange. The means of establishing identity depend partially on the crypto-graphic algorithm. If a public key algorithm is used, then secure authentication is almost always performed through a public key infrastructure (PKI). The next section presents a very brief introduction to PKI. If a shared key algorithm is used, the method of mutual authentication depends on whether both sides share a secret prior to initiation of key exchange. If the two sides do not share a secret, then a public key infrastructure can also be used to mutually authenticate. If the two sides share a secret, then authentication is often done by using an authentication, authorization, and accounting (AAA) protocol to a server that maintains a record of the node's identity and a preshared secret with the node, as discussed briefly in Chapter 1. Chapter 4 presents AAA protocols used in wireless Internet systems in more detail.

After authentication is complete, the actual key provisioning occurs. If a PKI is used for mutual authentication and a public key algorithm is used for cryptographic opera-tions, the other party's key is obtained as a side effect of authentication and no further operation is required. If provisioning with a shared secret is desired, two mechanisms are possible:

- Key exchange, where one or both sides encrypt a secret key and send it to the other party. Such key exchange requires either a preshared secret between the two parties, or that the two parties have each other's public keys. The use of an RSA key to encrypt a shared key, described above, is an example.
- Key derivation, in which both sides derive a shared secret from previously configured shared cryptographic material. Typically this mechanism is used for deriving shared keys from preshared secrets in AAA-based key derivation protocols, but it is also possible to derive a shared secret key from publicly known material using a Diffie–Hellman key exchange, which is discussed below.

3.5.1   Public key infrastructure (PKI)

Public key algorithms provide a convenient way for two parties to perform data origin authentication and confidentiality protection without much preconfiguration, but they have one drawback. When one party sends a public key to the other, how does the receiver know that the sender is, in fact, who they say they are? With algorithms based on the derivation of a shared key from a previously arranged shared secret, one party can verify the identity of the other by the fact that a shared key MAC can be verified and that MAC can only have been calculated by the party with which the key was shared in the past. But anybody can send a public key and claim it belongs to a node with a particular identity. What is needed is some kind of cryptographic proof of that identity.

As discussed in Chapter 1, systems that use public key algorithms have solved this problem by having the holder of the public key obtain a signed certificate containing the key holder's verifiable identity and the public key. The certificate is issued by a widely trusted third party, the *certification authority*. The public key holder sends the certificate to another party rather than the naked key. The certification authority signature on the certificate allows the receiver to verify that the public key holder is, in fact, who they say they are; provided of course the receiver has the certificate with public key from the certification authority.

If the receiver of a public key certificate doesn't have the certificate of the certification authority, checking a certificate may require the sender to provide a number of certificates for increasingly more broadly acknowledged certification authorities until the verifying party reaches a certificate for which it has a cached public key certificate. A node typically has a collection of such certificates for well-known certification authorities that are widely trusted. For example, a client attempting to verify the validity of the public key for a particular Web server may receive the following from the Web server:

- A public key certificate for the Web server itself signed by the certification authority of the service provider.
- A public key certificate for the service provider signed by a certification authority company that issues public key certificates broadly to companies on the Internet.

This process can continue for a number of certificates forming a chain. The certificate at the end of the certification chain is typically from a certification authority that issues certificates for a broad collection of sub-authorities worldwide. There are few such authorities, and Web browsers typically contain a collection of the root certificates for them. If the collection of well-known certificates on the host running the Web application contains the root certificate, the host can verify the chain of certificates, and thus the identity of the Web server.

A certification authority faces essentially the same problem of determining the identity for a public key holder when it issues a public key, but the constraints on the mechanisms for making the identity determination are somewhat looser. The identity determination may be done off line, for example over the telephone or in person, or the two parties may share a secret that they can use to authenticate each other. There is no requirement that the identity determination bootstrap from a public key algorithm in this case. In addition, a node receiving a collection of well-known certificates must verify the certificates before installing them. This can also be accomplished through some offline mechanism, or by bootstrapping from a well-known certificate that has been verified through an offline mechanism.

When a certificate's validity is being checked, in addition to checking the signature, a receiving node must determine whether the certificate is still valid. One check on validity is whether the certificate has expired. If the expiration date has been exceeded, the certificate is no longer valid. In addition, it is possible that a node may be compromised during the lifetime of its certificate. In that case, the certification authority revokes the certification of the node's public key. The certification authority keeps a *certificate revocation list* (CRL) indicating which certificates have been revoked even though their expiration dates indicate that they are still valid. A receiving node verifies that the certificate has not been revoked using a certificate revocation check. This check typically involves an online protocol check back to the certification authority. In some cases, the certification authority periodically distributes a list of revoked certificates against which a certificate can be checked. Online mechanisms are generally preferred because they limit the amount of data distributed (CRLs can run into the hundreds of megabytes) and because they ensure freshness of the information.

3.5.2   Diffie–Hellman key exchange

One prerequisite for provisioning of a shared key may seem to be either a preshared secret or a confidentiality-protected exchange of a shared secret between the two parties, for example using a public key to encrypt a shared key as was described above for RSA key exchange. While these methods are used quite widely, there is an algorithm that allows two sides to generate a shared secret using publicly exchanged information, without any confidentiality protection. The algorithm, known as Diffie–Hellman key exchange, is used in many protocols, including the Internet Key Exchange protocol (IKE) which is discussed in Chapter 6.

Diffie–Hellman key exchange is one of the oldest public key algorithms. It was published by Whitfield Diffie and Martin Hellman in 1976. The patent, issued in 1980 but now expired, also included Ralph Merkel as a co-author. Though that was the first public description of the protocol, in 1997 the British signals intelligence agency GCHQ revealed that the algorithm had been independently discovered by Malcolm J. Williamson some years prior to the publication by Diffie and Hellman. Diffie–Hellman key exchange is based on the difficulty of solving the discrete logarithm problem.

The algorithm starts by having the two parties who want to establish a shared key agree on two numbers, p and g, which they will use for further operations. The agreed numbers need not be kept secret. The basic restrictions on the two numbers are that p must be a large prime and g must be less than p, but other than that, there are a couple of qualifications that improve the security of the algorithm which are discussed below.

When both sides have established p and g, they then independently pick a secret random number of minimum 512 bits, S_1 and S_2. Each side then computes the following value:

$$T_i = g^{S_i} (\bmod p)$$

The two parties then send each other their T_i through an unencrypted channel.

When party i receives party j's T_j, it calculates the following value:

$$K_i = T_j^{S_i} (\bmod p)$$

Note that $K_i = K_j$ because:

$$K_i = T_j^{S_i} (\bmod p) = g^{S_j S_i} (\bmod p) = g^{S_i S_j} (\bmod p) = T_i^{S_j} (\bmod p) = K_j$$

For further security, p should be restricted such that $(p-1)/2$ is prime and $g^x \neq 1$ $(\bmod p)$ unless $x = 0 \ (\bmod p - 1)$.

While Diffie–Hellman is secure against passive attack, there is an active attack that can compromise the protocol. The protocol as described above has no provision for the two sides to prove their identities to each other. Consequently, it is possible for one party to inadvertently end up exchanging a key with an attacker acting as a man in the middle. The attacker maintains two keys, one for either side of the conversation, enabling the attacker to eavesdrop on conversations between the two sides. This problem occurs with unauthenticated Diffie–Hellman. It can be mitigated by having both parties authenticate themselves to each other, using a public key certificate or some other method, resulting in authenticated Diffie–Hellman.

A problem on computationally limited devices or in time-critical situations is that the computation of p and g is expensive. As a result, a protocol designer might build p and g into the key exchange protocol as constants. This allows an attacker to calculate a large table based on the constant p, causing the discrete logarithm problem to be broken for that p. Even though constructing such a table would be computationally expensive, it might be worth it because it would enable the attacker to break every key derived by the protocol, not just the key for a single user. Designing the key exchange protocol so that p is not constant for every protocol exchange ensures that an attacker can't construct such a table.

3.6   Summary

In this chapter, we discussed five classes of cryptographic algorithms that are important in wireless Internet security. The five classes of algorithms are message digests and cryptographic hash functions, shared key algorithms, public key algorithms, and secure key provisioning. For each class of algorithm, we discussed how the algorithm is used, different types of algorithms and some basic considerations on their security properties, and a representative algorithm.

For message digests, we examined the SHA-1 message digest function. Although theoretical results indicate that SHA-1 is more vulnerable to attack than previously thought, SHA-1 continues to be acceptable in some applications. For cryptographic hashes, we examined the HMAC keyed cryptographic hash function. HMAC is often paired with SHA-1 to perform the message digest prior to constructing the keyed hash. These two algorithms are widely used in Internet standards to calculate shared key message authentication codes for data origin authentication.

For shared key encryption algorithms, we discussed the two different types of shared key algorithms, block ciphers and stream ciphers. We briefly discussed attack characterization for block ciphers. Since stream ciphers are rarely used in Internet standards, we chose the new AES block cipher algorithm as the representative algorithm for discussion. AES is widely used in Internet standards.

For public key algorithms, we discussed the basic premise of public key cryptographic algorithms, a mathematical problem that is hard if the party does not possess the private key but easy if the party possesses the private key. We briefly described how public key algorithms are used for confidentiality protection and data origin authentication. We then discussed the most popular public key algorithm in Internet standards, RSA. RSA is based on the difficulty of factoring large primes.

For key provisioning, we first discussed the two basic ways in which two parties can end up with cryptographic material appropriate for a secure exchange: key exchange and key derivation. Key exchange involves one or both parties securely sending the other party a key, while key derivation involves both parties deriving a key from publicly and/or privately known material. Both techniques require the two parties to authenticate themselves to each other prior to the key provisioning. We then briefly presented public key infrastructure (PKI) as a way for two parties to obtain public keys certified by a third, mutually trusted party in order to satisfy the authentication requirement. Finally, we discussed the Diffie–Hellman key derivation algorithm, which allows two authenticated parties to derive a shared key using a public key algorithm.

4 Wireless IP network access control

Private wired access networks, such as a local area network deployed by a company typically require a user to be located in a particular physical facility for the terminal to access a physical data port. Physical access to the premises is therefore required for network access; so many companies depend on physical access control to regulate wired network access. With private wireless access networks on the other hand, access to a specific physical location is not always necessary for network access. The radio signals from wireless access points typically cover a roughly circular area of best reception. If part of the area of best reception lies outside of the physical space controlled by physical access control devices such as keys and key cards, it is possible for an attacker to gain access to an unprotected network by simply setting up a terminal in the parking lot. Modifying a wireless access deployment to confine wireless signal reception to the inside of a building is not usually possible. Radio propagation is difficult to control and some wireless signal is always available outside the area of best reception. An attacker could even take advantage of a weak signal to gain unauthorized access. Wireless private networks, unlike wired networks, therefore require some kind of network access control system to verify the identity of prospective network users.

Unlike private wired access networks, public wired access networks, such as dial up networks and DSL, have always required network access control systems because a user's location is not confined to a specific location with controlled physical access. A user of a dialup network may, for example, dial in from their home or from a hotel room. Public wireless access networks, like private wireless access networks, have the same characteristic, since the wireless link is available to anyone having a compatible terminal device within the geographic area within which the wireless link is deployed. In addition, again unlike private access networks, the network operator has a financial interest in controlling who gets to use the network. If the network operator's business is based on subscriptions, then only users with a subscription should be able to gain access to the network. If the network operator instead allows access to all who can provide billing information enabling the operator to charge for network usage, then only users who can provide such billing information should be able to gain access to the network. In either case, network access control is necessary to restrict access to subscribers or customers able to pay.

In this chapter, we discuss wireless network access control systems. Because of the need for network access control in public wired access networks, the currently deployed wireless network access control systems evolved from the existing public wired network access control systems. After a brief discussion of wireless network access usage models and their effect on access control, we apply the functional architectural approach to the network access control architecture represented by these deployed systems, including a threat analysis and development of functions and functional entities. We then discuss the two basic types of existing network access control systems that embody the architecture: subscription-based systems where a user must have an account with the network provider and hotspot systems where the access network provides access to any user indicating ability to pay. We review the protocols defining the network access control standards in currently deployed wireless Internet systems.

4.1   Wireless network access usage models

There are two different usage models for wireless networks:

- For laptops and some other devices, the user moves to a location such as a coffee shop or conference room, sets up the device, and then works for a while before shutting down the device and moving to a different location. This model is called the *nomadic* usage model. In this usage model, the user has no need for maintaining session continuity between locations. Note that this usage model is the same as a wired usage model for public access dialup or DSL networks, where the user connects to a dialup or DSL line in a hotel room, works for a while, then moves to another dial up or DSL connection at another physical location.
- For mobile phones and other handheld devices, the device is in use and a session is active while the user is moving around. Telephone calls and their attendant voice sessions are the canonical example of such a service, but data sessions in which the user is accessing location-based services are another possibility. In this model, application sessions stay active while the user is moving from one geographical location to another, a property called *session continuity*. Unlike the nomadic model, the *mobility* usage model has no wired equivalent, since it is not possible to use a wired connection while moving.

In theory, these two usages models are quite different, but, in practice, most wireless network designs attempt to accommodate both, since network operators are interested in maximizing the number of customers. While the threats to network access are largely the same for both usage models, the primary difference is the additional complexity that mobility injects into wireless access control design even for countering the same threats as in the nomadic model. When mobility must be supported, the terminal will be breaking and re-establishing link layer and IP layer connectivity as it moves in and out of range of wireless access points, while maintaining its session.

The security state established at the initial access point when network access is granted forms a security association between the access point and the terminal. The security association contains keys providing confidentiality and data origin authentication on link layer frames exchanged between the access point and the terminal. If the terminal moves its link connection to a new access point, the security association must be re-established on the new access point. In the nomadic usage model, no session continuity is required between moves, so the security association can be re-established from scratch. In the mobility usage model, on the other hand, the security association must be established quickly, since real-time traffic such as voice may experience unacceptable performance if delays occur. Usually the initial network access control procedure requires more time than is acceptable for maintaining a real-time session, so additional modifications are required. Much of the complexity in network access control systems results from having to accommodate the mobile usage model. For this reason, we primarily concentrate in this chapter on the network access control architecture for the nomadic usage model.

4.2 Threats to wireless network access

The following are a list of common threats to wireless network access:

- The base threat that wireless network access systems must deter is unauthorized network access. If the network requires users to have an account, unauthorized network access occurs when a terminal cannot prove that it has an account. If the network allows access to any terminal that can prove ability to pay, unauthorized network access occurs when the terminal cannot provide proof of ability to pay, yet it nevertheless manages to obtain access to the network.
- If the network offers different kinds of service, another threat occurs if a terminal authorized for a lower class of service is able to obtain access to a higher class of service without authorization.
- Once an authorized terminal has gained access to its allowable class of service, an attacker could attempt to steal the terminal's session and thereby obtain network access even though the attacker is not authorized, or the attacker could attempt to upgrade service without authorization by hijacking the authorized terminal's session. In order to do this, the attacker would have to spoof the authorized terminal, so this is a kind of spoofing attack.
- An eavesdropper could intercept the terminal's traffic on the air and extract information from the traffic that could harm the terminal's user, or drop the traffic denying service to the terminal.
- While most of the threats above derive from an attacking terminal, there is also a threat from the network side. An attacker could set up a rogue access point or base station, and thereby spoof a legitimate terminal into connecting. The rogue access point could then examine traffic and conduct a variety of attacks.
- In access networks that support the mobility usage model, a threat related to session stealing occurs when an authorized terminal is handing over from an access point on which it is fully authenticated to another on which it is not yet authenticated. If the handover protocol security is not properly designed, an attacker could take over a victim's session at the old access point for a period of time after the victim has moved, or the attacker could conversely hijack the session on the new access point while the victim is attempting to authenticate and set up the new session. This threat demands careful attention to security around access point handover, and the corresponding operation at the IP level, handover between IP subnets.

The security services used to mitigate the above attacks are the following:

- Identity management in the form of authentication and authorization addresses threats involving unauthorized entities. Identity verification of the terminal by the network mitigates the threat of unauthorized network access service or access to a service level above the authorized level. Identity verification of the network by the terminal mitigates the threat of a rogue access point or base station attempting to steal the terminal's traffic.
- Key management establishes keys for cryptographic algorithms used in further security services. The initial authentication and authorization process includes key management to ensure that the keys are tied to a verified terminal or network identity.
- Data origin authentication using the keys established during authentication and authorization mitigates the threat of session stealing, both on the current access point and during handover.
- Confidentiality protection using the keys established during authentication and authorization mitigates the threat of eavesdropping.

In the next section, we discuss the functional architecture, and show how these security services translate into specific functions and functional entities in the network access control architecture.

4.3   Functional architecture for network access control

The following functional entities model those in existing, deployed network access control systems:

- The Supplicant is the authentication and authorization entity on the wireless terminal that is requesting network access.
- The Authenticator is the authentication and authorization entity in the wired access network at the first point of contact between the wireless terminal and the wired network (typically at the wireless access point/base station or the first hop access router) that is performing access control.
- The Account Authority is the root of trust between the Supplicant and Authenticator that can authenticate the Supplicant's credentials and determine for what type of service the Supplicant is authorized. The Account Authority may also handle key management. The Account Authority is usually a server somewhere in the access network or in the Internet.

This description contains no specific binding of the entities to specific network devices. Later in the chapter, we discuss such binding for two specific examples of network access control systems.

4.3.1   Functional architecture and interfaces

Figure 4.1 shows the overall functional architecture for the network access authentication system including the functions, functional entities, and interfaces between functional

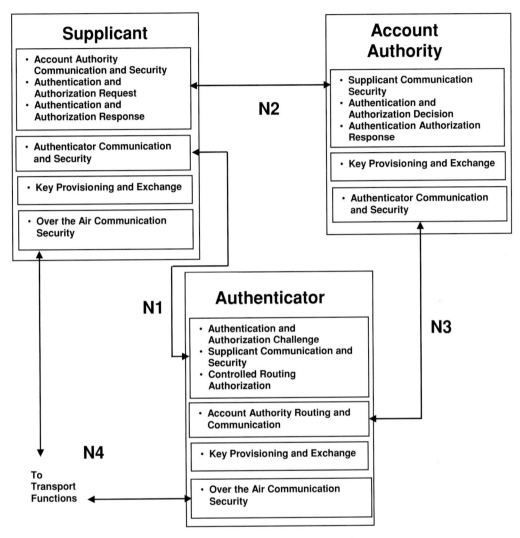

Figure 4.1 Functional architecture for network access control system

entities. Interfaces between functions internal to a particular functional entity are programmatic. The arrows between the network entities indicate network interfaces, and they terminate on the collections of functions with the functional entities that are connected.

There are four open network interfaces in the network access authentication architecture:

- N1 – This interface connects the Authenticator and the Supplicant. This interface involves functions that allow the Authenticator to challenge the Supplicant to provide authentication and prove authorization, that confirm session key possession or allow session key provisioning after network access has been granted, and that provide security over the air between the Authenticator and Supplicant during initial network access control. This interface does not include any functions for securing traffic between the terminal and network after network access control has completed.

- N2 – An end-to-end interface connecting the Supplicant with the Account Authority. Traffic between the Supplicant and the Account Authority flows through the Authenticator, although the Authenticator does no more than route the traffic to the proper Account Authority in a fashion roughly similar to an overlay network. This interface involves functions that allow the Supplicant to provide credentials in a secure fashion to the Account Authority proving authentication and authorization for services, and allowing the Account Authority to reply to the Supplicant about the outcome of the authentication and authorization check, including key provisioning if the Account Authority provisions session keys to the Supplicant directly. In addition, the Supplicant can check the identity of the Account Authority. Functions in the interface also allow the Account Authority to check the Supplicant's credentials through a programmatic interface.

- N3 – This interface connects the Authenticator and Account Authority. Functions in this interface allow secure communication between the Authenticator and the Account Authority, including the secure tunneling and routing of N2 messages between the Supplicant and the Account Authority. In addition, the interface includes functions that allow the Account Authority to indicate whether access has been granted to the Supplicant and to provision session keys on the Authenticator, if the Account Authority is responsible for provisioning keys on the Authenticator.

- N4 – Once network access authentication is complete, functions in this interface secure the on-going user data traffic between the terminal and the network. This is not part of the network access control system but is part of the security architecture. We show this interface between the Supplicant and the Authenticator, though, strictly speaking, the interface itself is between the terminal and access point which are not network access control functional entities. The terminal and access point internally call upon Supplicant and Authenticator functions, respectively, to carry out the security functions.

4.3.2   Supplicant functions

Table 4.1 contains a list of Supplicant functions together with the security services they provide, the parameters for the functions, and the objects returned by the functions. The functions can be tied back to the threats through the security services and the discussion above. The following subsections describe the functions in more detail.

Authenticator Communication and Security function

The Authenticator Communication and Security function conducts the network access control communication between the Supplicant and Authenticator. If there is any network access control-specific security on the communication, it is maintained by this function. If sending, the parameters are a formatted message to the Account Authority, the Account Authority routing information, the Authenticator routing information, and any link specific parameters including security parameters. If receiving, the parameters are a

Table 4.1 Functions, parameters, and results for the Supplicant

Function	Security services	Parameters	Return
Authenticator Communication and Security	– Manage secure communication with the Account Authority for identity management and key management during network access control	– Formatted message to the Account Authority for authentication and authorization if sending – Account Authority identity for routing if sending – Authenticator identity for routing if sending – Formatted message from the Authenticator if receiving – Other link-specific parameters including any keying material needed to secure the Supplicant to Authenticator network access control transaction	– Formatted message to the Authenticator if sending – A formatted message from the Account Authority if receiving for use in the Account Authority Communication and Security function
Account Authority Communication and Security	– Data origin authentication and confidentiality protection on communication with the Account Authority	– Formatted message to/from the Account Authority – Account Authority identity – Keying material for securing/verifying message for/from Account Authority	– Confidentiality and data origin authentication protected (secured) message to the Account Authority if sending – Verified clear text message from Account Authority if receiving
Authentication and Authorization Request	– Provide terminal identity authentication – Request network identity verification – Request authorization for service	– Account identity – Credentials proving identity – Service request – Credentials proving authorization for service request	– Clear text message to the Account Authority requesting authentication and authorization
Authentication and Authorization Response	– Verify network identity – Confirm authorization for network access – Manage remote provisioning of session keys for wireless link	– Clear text message from the Account Authority containing response to authentication and authorization request and cryptoparameters for key provisioning (if any)	– Yes/no decision about whether the Account Authority has approved network access and the requested service type – Session keys for session if yes/no decision is positive
Key Provisioning and/or Exchange	– Local provisioning of session keys for wireless link traffic between the terminal and the network	– Cryptoparameters for session key generation (if any) with Account Authority and Authenticator	– Session keys for confidentiality protection and data origin authentication shared with Account Authority and Authenticator
Over the Air Communication Security	– Data origin authentication and confidentiality protection on user data traffic over the wireless link after network access control is complete	– Session keys for securing/verifying message with wireless access point – Message to/from wireless access point that needs protection or verification/decryption	– Confidentiality and data origin authentication protected (secured) message to the wireless access pont or clear text message from the wireless access point

formatted message from the Authenticator and any link specific parameters. The function returns a formatted and secured message to the Authenticator if sending and a formatted and secured message from the Account Authority if receiving.

Account Authority Communication and Security function

The Account Authority Communication and Security function handles end-to-end communication and security with the Account Authority. Using the end-to-end keying material shared with the Account Authority, this function takes a clear text message to the Account Authority and the Account Authority identity and produces a message with confidentiality and data origin authentication protection, formatted properly for the end-to-end communication protocol with the Account Authority. The function also processes return messages from the Account Authority, taking a secured message and the Account Authority identity, and returning a formatted clear text message.

Authentication and Authorization Request function

In response to an Authentication and Authorization Challenge from the Authenticator or autonomously (depending on the particular system and operation), an Authentication and Authorization Request message is formulated to the Account Authority for transmission via the Authenticator. The message contains the account identity, service request and credentials proving account identity and authorization for service. The function returns a properly formatted clear text message to the Account Authority.

Authentication and Authorization Response function

The Authentication and Authorization Response function takes as input the long-term keying material shared with the Account Authority and a confidentiality and data origin protected message from the Account Authority containing the cryptoparameters for session key generation. If the message indicates that the Account Authority has approved the request for network access and authorized the requested service, the function invokes the Key Provisioning and Exchange function to set up session keys for the session between the Supplicant, Authenticator, and Account Authority. The function returns an indication of whether the Account Authority has approved the request for network access and the service type. This indication can be used by other terminal subsystems to determine whether to start sending user packets across the wireless link or whether another attempt at authentication and authorization, perhaps with a different requested service type, is necessary.

Key Provisioning and/or Exchange function

The Key Provisioning and/or Exchange function generates or provisions session keys shared with the Account Authority. The exact nature of the function depends on the key provisioning algorithm. If the Account Authority and Supplicant share a long-term secret, then the session keys are generated using the long-term secret and material from the message exchange. If the Account Authority provides Diffie–Hellman parameters, then a Diffie–Hellman exchange is used to generate a master key from which session keys are generated. If the Account Authority has encrypted a shared key using the

Supplicant's RSA or other public key, then the master shared key is decrypted and used to generate session keys.

Over the Air Communication security function

The Over the Air Communication Security function is invoked by the wireless link transmission functions after network access has been established in order to provide data origin authentication and confidentiality protection on user data frames sent between the terminal and the wireless access point. The function is also responsible for invoking the Authentication and Authorization Request message when new session keys need to be generated, for example, if the session keys time out or if the terminal hands over to a new access point. Note that this function is not part of the Supplicant network access control functional entity, since it is invoked after network access control is complete.

4.3.3   Authenticator functions

Table 4.2 contains a list of Authenticator functions together with the security services they provide, the parameters for the functions, and the objects returned by the functions. The following subsections describe the functions in more detail.

Supplicant Communication and Security function

The Supplicant Communication and Security function acts as an intermediary for network access control communication between the Supplicant and the Account Authority. It conducts relay communication over the wireless link. If there is any network access control-specific security on the communication, it is maintained by this function. If sending, the parameters are a formatted message from the Account Authority to the Supplicant, the Supplicant identity, and any link specific parameters including security parameters. If receiving, the parameters are a formatted message from the Supplicant to the Account Authority and any link-specific parameters. The function returns a formatted and secured message to the Supplicant if sending and a formatted and secured message from the Supplicant to the Account Authority if receiving.

Authentication and Authorization Challenge function

Upon detection of an attempt by an unauthenticated terminal to access the network, the Authentication and Authorization Challenge function is called. It issues a challenge requesting that the terminal initiate authentication and authorization. The function takes parameters indicating the Authenticator's identity and the Supplicant's identity and returns a formatted message to the Supplicant issuing the challenge. This function also periodically reissues a challenge if session keys time out.

Account Authority Routing and Communication function

The Account Authority Routing and Communication function is responsible for securely relaying Supplicant traffic to/from the Account Authority and for handling messages exchanged directly between the Account Authority and the Authenticator. The parameters are routing information for the Account Authority with which the communication is

Table 4.2 Functions, parameters, and results for the Authenticator

Function	Security services	Parameters	Return
Supplicant Communication and Security	– Manage secure communication with the Supplicant for identity management and key management during network access control	– Supplicant identity – A formatted message to/from the Account Authority from/to the Supplicant – Other link-specific parameters including any keying material needed to secure the Supplicant to Authenticator network access control transaction	– Formatted message to the Supplicant if sending – Formatted message from the Supplicant to/from the Account Authority or Supplicant – Account authority identity for routing if receiving
Authentication and Authorization Challenge	– Initiate network access control	– Authenticator identity – Supplicant identity	– Formatted message to the Supplicant
Account Authority Routing and Communication	– Manage routing of messages between the Supplicant and Account Authority during network access control – Manage secure communication with the Account Authority for network access control decision and key provisioning	– Account Authority routing information – Formatted message to/from the Supplicant from/to the Account Authority – Long-term keying material shared with Account Authority	– If sending, a formatted and secured message to the Account Authority including the Supplicant's request – If receiving, yes/no result from Account Authority allowing/denying Supplicant access – If receiving, a formatted message from the Account Authority to the Supplicant
Controlled Routing Authorization	– Confirm authorization for network access – Initiate authorized network access for the terminal – Manage remote provisioning of session keys for wireless link	– Permission status from Account Authority	– <none >
Key Provisioning and Exchange	– Local provisioning of session keys for wireless link traffic between the terminal and the network	– Cryptomaterial from Account Authority for generating master session keys with Supplicant	– Session keys shared with Supplicant for confidentiality and data origin authentication protection of over the air communications
Over the Air Communication Security	– Data origin authentication and confidentiality protection on user data traffic over the wireless link after network access control is complete	– Session keys shared with terminal for confidentiality and data origin authentication protection – Message to/from terminal that needs protection or verification/decryption	– Confidentiality and data origin authentication protected (secured) message to the terminal or clear text message from the terminal

to occur, the long-term keying material shared with the Account Authority for securing the message, and a secured (confidentiality and data origin authentication protected) message from the Supplicant to the Account Authority or from the Account Authority to the Supplicant. If relaying a message to the Supplicant, the function returns a yes/no result from the Account Authority confirming or denying access, and a formatted and secured reply message from the Accounting Authority to the Supplicant for input into the Supplicant Communication and Security function. If relaying to the Account Authority, the function returns a formatted and secured message to the Account Authority from the Supplicant.

Controlled Routing Authorization function

The Controlled Routing Authorization function is responsible for opening up routing to/from the Internet for the Supplicant. The parameter is the Account Authority's permission for denial of service.

Key Provisioning and/or Exchange function

The Key Provisioning and Exchange function takes cryptomaterial sent by the Account Authority for exchanging or generating shared keys with the Supplicant and generates session keys for the Supplicant. Depending on the particular key provisioning algorithm, the cryptomaterial might be a pairwise master key shared indirectly between the Authenticator and Supplicant via the Account Authority, or it might be some public keying material such as Diffie–Hellman parameters. Further communication with the Supplicant may be necessary if the Supplicant and Authenticator directly exchange keys.

Over the Air Communication Security function

The Over the Air Communication Security function is the matching function to that on the terminal. It handles data origin authentication and confidentiality protection of frames communicated between the wireless access point and the terminal after network access authentication has completed. Parameters are the session keys for data origin authentication and confidentiality protection shared with the terminal and a message to be protected. The return is the protected message or the clear text message. As with the terminal side, this function is not part of the Authenticator network access control functional entity.

4.3.4 Account Authority functions

Table 4.3 contains a list of Account Authority functions together with the security services they provide, the parameters for the functions, and the objects returned by the functions. The following subsections describe the functions in more detail.

Supplicant Communication and Security function

The Supplicant Communication and Security function handles end-to-end secure communication between the Account Authority and the Supplicant. Parameters are the long-term keying material shared with the Supplicant, and a message to or from the

Table 4.3 Functions, parameters, and results for the Account Authority

Function	Security services	Parameters	Return
Supplicant Communication and Security	– Manage secure communication with the Supplicant for identity management and key management during network access control	– Long-term keying material shared with Supplicant – Message to/from Supplicant to be protected or verified/decrypted	– Message to/from Supplicant that is protected or verified clear text
Authenticator Communication and Security	– Manage secure communication with the Authenticator for identity management and key management during network access control	– Long-term keying material shared with Authenticator – Message to/from Authenticator to be protected or verified/decrypted	– Message to/from Authenticator that is protected or verified clear text
Authentication and Authorization Decision	– Identity verification – Terminal authorization verification	– Verified clear text authentication and authorization request message from Supplicant	– Yes/no decision whether the Supplicant is authenticated and authorized for requested service
Authentication and Authorization Response	– Provide network identity authorization to Supplicant – Provide authorization decision to Supplicant and Authenticator – Manage remote provisioning of session keys to Supplicant and/or Authenticator	– Authentication and authorization decision outcome – Authenticator routing information – Supplicant identity – Session keys for confidentiality protection and data origin authentication shared with Account Authority and Authenticator	– Response message to Authenticator indicating authentication and authorization decision including key provisioning
Key Provisioning and/or Exchange	– Local provisioning of session keys for wireless link traffic between the terminal and the network	– Keying material for provisioning the Supplicant and Authenticator	– Session keys for confidentiality protection and data origin authentication shared with Account Authority and Authenticator

Supplicant that either is in clear text and needs protection or is protected and needs verification and decryption. The return is the message either protected or in clear text, depending on the input parameter.

Authenticator Communication and Security function

The Authenticator Communication and Security function performs the same operation for communication between the Account Authority and the Authenticator. Parameters are the long-term keying material shared with the Authenticator and a message to or from the Authenticator, possibly containing an embedded protected message to the Supplicant, that either is in clear text and needs protection or is protected and needs verification and decryption. The return is the message either protected for the Authenticator or with the part from the Authenticator in clear text. Any embedded protected message from the Supplicant is included in the return and must be handed off to the Supplicant Communication and Security function after completion of the function.

Authentication and Authorization Decision function

The Authentication and Authorization Decision function takes a clear text message from the Supplicant requesting authentication and authorization for a particular service type and returns a yes/no decision about whether the supplicant is authenticated and authorized for the service.

Authentication and Authorization Response function

The Authentication and Authorization Response function takes the decision outcome from the Authentication and Authorization Decision function, the Authenticator identity, and the Supplicant identity, and session keys for the Supplicant and Authenticator and formulates a protected response to the Supplicant via the Authenticator indicating the decision, and a response to the Authenticator indicating the decision. The responses also contain any session keys.

Key Provisioning and/or Exchange function

The Key Provisioning and Exchange function includes keying material for provisioning the Supplicant and Authenticator with session keys for data origin authentication and confidentiality protection and returns the session keys.

4.3.5   Additional design requirements

Besides the design requirements established by the threats, two additional design requirements that are often important for network access control systems are support for roaming and minimal cryptoboundaries for provisioned or derived keys. Roaming is important in systems where users are required to have an account. By supporting roaming, the access network provides service to traveling users who do not have an account with the local access network provider but who do have access with a home network provider in another geographical location. Roaming is common in wired dialup networks too. The cryptoboundary concept was briefly mentioned in Chapter 1. Cryptoboundaries are useful for thinking about how to design key distribution, to reduce the number of places from which attacks can be mounted.

Roaming and network access control

Many wireless service providers establish service contracts with their users, which have a duration longer than a single network session. In the process of setting up the service contract, the service provider provisions security credentials on the device prior to the first network access and establishes a service profile on an AAA server for the user and terminal. The service profile allows the network access control system to authenticate the terminal's credentials when the terminal attempts to access the network. The access control procedure is very straightforward if the terminal is attempting to access a network owned by the service provider with which the user has an account. This is the home network service provider.

Users that travel frequently may want wireless network access when they are not within the coverage area of their home service provider. In these situations, the user and terminal are said to be *roaming*. The local access network service provider then needs to conduct network access control for the terminal prior to providing Internet service. For networks where the service provider has a longer-term contract with its customers, the home and local network provider typically have a business relationship allowing mutual roaming among their customers. The local network provider contacts the home network service provider to ensure that the terminal is properly authenticated and authorized for service. This transaction can be initiated by the terminal or the network depending on the particular AAA architecture in use, More details are discussed below.

The terminal identifies itself to the local network using a *Network Access Identifier* (NAI). RFC 4282 (RFC 4282, 2005) describes the NAI. The NAI is structured like an email address. For example, the NAI for Bob Smith whose home access network provider is Stanford University might be "bob.smith@stanford.edu". The NAI allows the local access network to route requests for access from roaming terminals back to the home network. Intermediate networks between the local access network and the home network also use the NAI for routing. The home network uses the NAI to identify the service profile for the terminal's user, and – combined with data origin authentication on the message – allows the home network to authenticate and check authorization for the user and terminal.

The cryptoboundary concept

An important side effect of AAA is often the provisioning of a shared key in the access network and terminal. The shared key is established for data origin authentication and confidentiality protection of data traffic over the wireless link. During key provisioning, the entities that have access to the shared secret need to be strictly controlled. Unauthorized access to a shared secret during provisioning can be controlled by providing confidentiality protection on messages between the key supplier and a client, if the key is generated by one party to the conversation and provisioned over the network to the other. An even more effective way to limit unauthorized access is to derive the session keys in parallel on the wireless terminal and in the network using an algorithmic derivation from a preshared secret, as described in Chapter 1. No confidential material is distributed over the network. Parties that need access to the key but do not have access to the long-term secret can be provisioned with the key over the network, as long as confidentiality and data origin authentication are maintained on the network transaction. Periodically, the two sides re-derive the session keys to reduce exposure to key compromise.

Even if proper security measures are taken to limit unauthorized access during provisioning, one party in the conversation could still become compromised at some point after the key has been provisioned. Re-deriving the key periodically helps reduce the time period in which compromise goes undetected, since the time period will be limited by the validity duration of the session key, unless, of course, the compromised party has access to the long-term secret. The renewal period can be related to the probability of brute-force compromise, or the period can be set based on system management considerations if the key size is large and therefore the probability of brute-force compromise relatively small.

Reducing the number of entities that have access to a provisioned, shared key can limit the size of the potential target population for an attacker. Fewer entities with access to the shared key mean fewer targets for an attacker. The number of entities that have access to a shared key is called the *cryptoboundary* of the key. The cryptoboundary is a useful concept for limiting the extent of a potential key compromise. Ideally, in the case of a shared key architecture, the cryptoboundary encompasses only the two parties having access to the key. In public key architectures, the cryptoboundary is usually restricted to a single node, the node that generated and possesses the private key corresponding to the public key. Fewer nodes having access to a key results in less damage should the key be compromised.

4.3.6 Taxonomy of deployed systems

The functional architecture described in the above sections could lead to a variety of possible system types depending on choice of cryptosystems and protocols for network access authentication and authorization, data origin authentication, and confidentiality protection. We examine two approaches here that have been used in 802.11/WiFi network deployments. These approaches resulted from heavy constraints during the standardization and deployment process to make network access control backward compatible with existing, widely deployed Web-based technology or with dialup network access control systems with deployed AAA servers and protocols. These two approaches are:

- Subscription-based Approach – The network access control support is provided by the wireless link protocol, together with the same AAA protocols and backend technology used originally in dialup systems. This architecture is usually deployed by enterprise networks and public access network where the terminal's user typically – although not always – has a subscription with the network provider. The terminal may prove its identity with a login/password, a public key certificate, or a shared key MAC. The AAA server in the home network maintains an account profile, which may contain a preshared secret with the terminal. The terminal requires special AAA software to conduct the AAA transaction. Depending on the wireless link protocol authentication procedure, the access point or base station may or may not authenticate with the terminal.

- Hotspot Design – The hotspot design is used by walk-up networks, called *hotspots*, in which the user need not have a subscription with a service provider (though some hotspot networks also support subscriptions). These networks are primarily concerned with securely setting up accounting on a per use basis so that the user is charged for network access. The network access control transaction is conducted through a Web page using a secure HTTP connection. Security is not provided over the wireless connection after the terminal has obtained network access. Users are expected to provide their own data origin authentication and confidentiality protection over the wireless link. When such protections are available (and they really always ought to be), they are usually provided through establishing a Virtual Private Network (VPN) between the user's terminal and some wired network, often a home corporate network or VPN service provider. Traffic between the terminal and the VPN server is protected with data origin authentication and confidentiality protection, thereby protecting traffic over the wireless link.

Table 4.4 provides a mapping between the functional architecture developed in the previous sections and the two deployed system designs. With the exception of the Public Access Control Gateway, the functional elements and protocols used in the hotspot design are all not specific to network access control systems. They are mostly reused from other systems, such as Web browsing and e-commerce. This is actually by design, since hotspot networks must accommodate any terminal that walks up to the network, including terminals that do not have any specialized AAA software or hardware. In contrast, the AAA server-based design has functional elements and protocols dedicated to network access control. In some cases, the terminal may even have specialized hardware, such as a secure smart card, for assisting in network access control operations.

In the following sections, we discuss these two approaches. Since the subscription design is specific to the particular wireless link protocol, we use 802.11 as an example. The AAA support provided by other wireless link protocols naturally differs in various ways. The protocols involved in the subscription design are specific to network access control, and we briefly examine those used in 802.11 network access control. The hotspot design is very generic and independent of the wireless link layer protocol. It can be implemented entirely above the IP layer. The protocols involved in implementing the hotspot design are very general and not specific to network access control. This is a natural result of the basic design goal of the system. If a hotspot is to be used by any terminal, then specialized network access control software cannot be a prerequisite.

4.4   Subscription-based design

In the subscription-based design, network access is authorized by an AAA server in the home network of the terminal. Naturally, for network access to succeed, the user must

Table 4.4 Mapping of interfaces to protocols for example deployed system designs

	Functional elements			Protocols on interfaces			
	Supplicant	Authenticator	Account Authority	N1	N2	N3	N4
AAA Server- Based Design	Terminal Layer 2	Wireless Access Point Network Access Server (NAS)	Home Network AAA Server	Link Specific (e.g. 802.1x)	EAP or other end to end	Radius	Link Specific (e.g. 802.11)
Hotspot Design	Terminal Web Browser	Public Access Control Gateway HTTP Proxy	Credit Card Network Server	HTTP/HTTPS	<None>	Ecommerce Protocol	IKE and IPsec (optional)

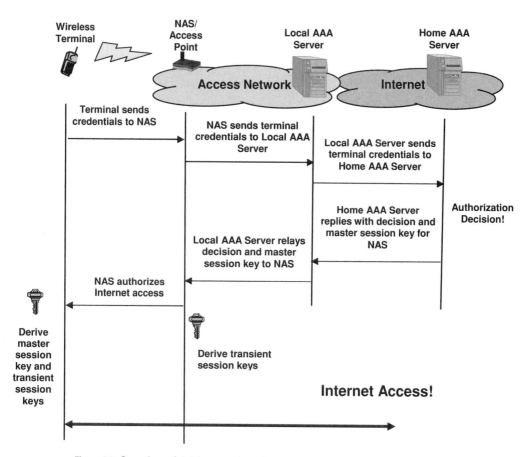

Figure 4.2 Overview of AAA server-based network access control

have an account with a service provider that has a roaming relationship with the local access network provider. Figure 4.2 contains an overview of the protocol. The terminal conducts an AAA session with the AAA server in the home network prior to obtaining an IP address. The network access server (NAS) in the access network routes the AAA requests from the terminal but does not otherwise participate in the exchange between the terminal and the AAA server. If the terminal is a roaming customer, the NAS may route the traffic through the local AAA server, as shown in the figure.

The terminal first sends its credentials to the NAS over the wireless link. The NAS forwards the credentials to the local AAA server, which routes them to the home AAA server. The home AAA server checks the terminal's credentials, sets up accounting if necessary, and replies back through the local AAA server and NAS, authorizing or denying access. The NAS then forwards the reply to the terminal. Before access, the terminal is pre-provisioned with a secret shared with the home AAA server. The home AAA and the terminal derive a master session key from the preshared secret in parallel. The AAA server securely provisions the NAS with the master session key over a confidential channel. The terminal and NAS use the master session key to generate further transient session keys, which are used for various applications, such as data origin authentication and confidentiality protection of traffic over the wireless link.

Historically, the AAA server-based design was used in wired dialup networks, in which a modem connected a PC to an access network over a circuit-switched telephone line. The PC utilized a serial line protocol such as Point to Point Protocol (PPP), described in RFC 1548 (RFC 1548, 1993), to impose a frame structure on the serial link across the telephone line. Prior to gaining network access, the PC conducted an authentication exchange with the AAA server in the access network via the NAS located in the modem pool at the network end of the phone connection. PPP supported a variety of protocols for doing the authentication exchange.

As adapted to wireless network access control, the role of PPP is played by a wireless-link specific protocol that runs between the NAS and the terminal. This protocol encapsulates the authentication exchange over the wireless link rather than using IP because, prior to the success of the authentication and authorization exchange with the AAA server, the terminal has no IP address. An example of a link-specific network access protocol is the 802.1x protocol (802.1x, 2004) for 802.11 and other 802-based link layers, which we examine in further detail later in this section.

Wireless systems use many of the same protocols between the terminal and the AAA server that are used over PPP in serial line systems. One of the most widely used is Extensible Authentication Protocol (EAP) (RFC 3748, 2004). As its name implies, EAP is extensible and supports a variety of different kinds of authentication methods between the client and the server. This allows wireless network service providers to configure their authentication and authorization systems with a variety of different technologies, depending on their existing infrastructure.

The final protocol piece in wireless network access control systems is the protocol between the NAS and the AAA server. In most deployments, the Radius protocol is used for this purpose (RFC 2865, 2000). Radius allows the NAS and the AAA server to communicate a variety of information using attribute/value pairs. The types of attribute/value pairs are standardized, although there is an attribute/value pair format that allows vendors to introduce their own, nonstandardized types.

The next three subsections discuss the protocols from the IEEE 802.1x/802.11–2007 network access authentication (802.11, 2007) (802.1x, 2004). This system is widely deployed for enterprise wireless access in WiFi (802.11) networks, and in some cases for public access WiFi networks as well. The protocols in the system are:

- EAP for end-to-end authentication between the terminal and the AAA server;
- 802.1x to carry EAP over the wireless LAN link (EAPoL), between the terminal and the NAS;
- Radius to transport EAP between the NAS and the AAA server, and to carry provisioned keys and other information between the NAS and AAA server.

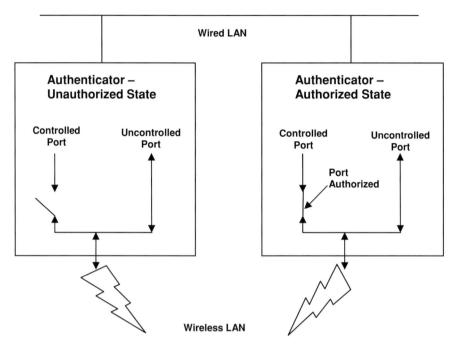

Figure 4.3 802.1x authenticator architecture

4.4.1   802.1x/EAPoL

The 802.1x protocol is used between the wireless terminal and the 802.11 access point to transport EAP messages. As shown in Figure 4.3, the 802.1x standard specifies an authentication architecture in which the Authenticator in the 802.11 access point offers two different ports to the Supplicant in the terminal:

- an *uncontrolled* port over which only authentication traffic can flow;
- a *controlled* port over which arbitrary types of traffic can flow but only after authentication is complete.

The Authenticator only unblocks the controlled port after it is informed by the AAA server that a successful authentication transaction has been conducted by the terminal. The closing of the controlled port provides the terminal with access to the wired LAN and, through the local access router, to the Internet.

802.1x uses EAP over LAN (EAPoL) to transport EAP between the terminal and the access point. EAPoL is a special kind of Ethernet frame that encapsulates the EAP protocol. Prior to performing the EAP transaction, the terminal must associate with the access point. The terminal addresses EAPoL frames to the access point's link layer address.

Figure 4.4 illustrates the frame format for EAPoL. The fields have the following definitions:

- Port Access Entity Type: Ethernet frame type for EAPoL, 0x888e
- Version – protocol version, 0x02 for the 2004, 0x01 for the 2001 (original) version

Table 4.5 EAPOL frame type descriptions

Name	Value	Description
EAPOL-Packet	0x0000	Encapsulates all EAP protocol data. Forwarded by the Authenticator between the terminal and the AAA server
EAPOL-Start	0x0001	Sent by the Supplicant to initiate an EAP exchange
EAPOL-Logoff	0x0002	Sent by the Supplicant to terminate use of the controlled port
EAPOL-Key	0x0003	Transmits global key information from the Authenticator to Supplicant if supported
EAPOL-Encapsulated-ASF-Alert	0x0004	Used for network management

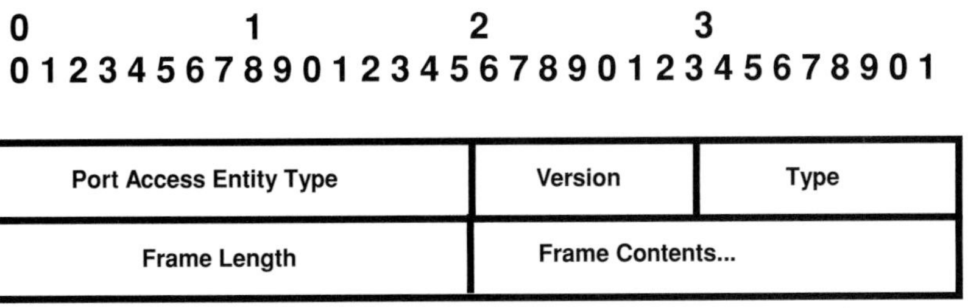

```
0                   1                   2                   3
0 1 2 3 4 5 6 7 8 9 0 1 2 3 4 5 6 7 8 9 0 1 2 3 4 5 6 7 8 9 0 1
```

Port Access Entity Type	Version	Type
Frame Length	Frame Contents...	

Figure 4.4 EAPOL frame format

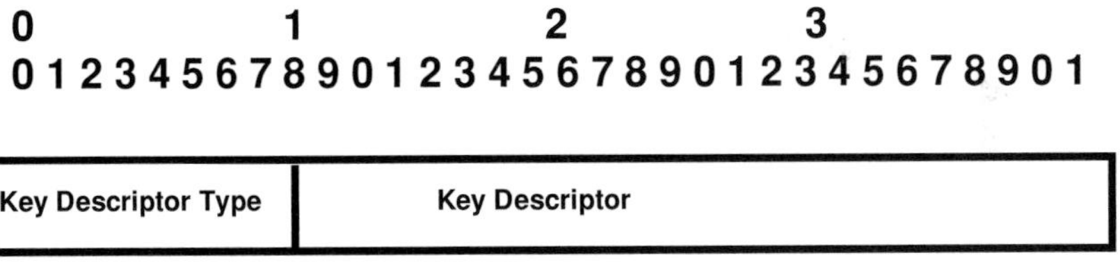

```
0                   1                   2                   3
0 1 2 3 4 5 6 7 8 9 0 1 2 3 4 5 6 7 8 9 0 1 2 3 4 5 6 7 8 9 0 1
```

Key Descriptor Type	Key Descriptor

Figure 4.5 EAPOL-KEY frame body format

- Type – the EAP frame type, see Table 4.5
- Frame Length – the length of the Frame Contents field, in bytes
- Frame Contents – the EAP protocol data

Note that EAPoL itself does not provide support for data origin authentication or confidentiality and anti-replay protection. On 802.11, these services are typically provided end to end, between the Authentication Server and the Supplicant in the terminal, and not between the Authenticator in the access point and the Supplicant.

If the EAPoL frame is of type EAPOL-Key, the frame body contains a message formatted as shown in Figure 4.5. For 802.11 key provisioning, the value of the Key Descriptor Type field is 0x02. The Key Descriptor field is described in the next section, since this message is used together with EAP for key provisioning.

Table 4.6 EAP message type descriptions

Name	Value	Description
EAP Request	0x01	Encapsulates a network access control request
EAP Response	0x02	Encapsulates a reply to a network access response
EAP Success	0x03	Sent by the Authenticator to the Supplicant to indicate successful authentication
EAP Failure	0x04	Sent by the Authenticator to the Supplicant to indicate authentication failed

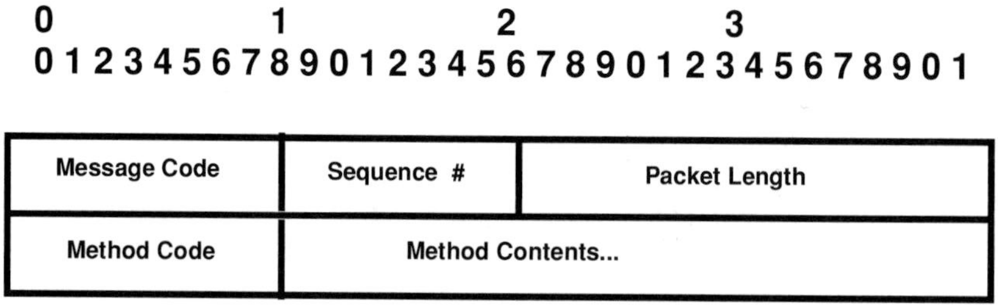

```
 0                   1                   2                   3
 0 1 2 3 4 5 6 7 8 9 0 1 2 3 4 5 6 7 8 9 0 1 2 3 4 5 6 7 8 9 0 1
```

Message Code	Sequence #	Packet Length
Method Code	Method Contents...	

Figure 4.6 EAP packet format

4.4.2   EAP

EAP is an encapsulation mechanism and request/response protocol for transporting requests for network access authentication and responses to those requests between the terminal and the network. EAP itself contains no support for network access operations. Instead, different EAP methods have been defined that implement user/terminal identity authentication, network access authorization, and key configuration upon initial network entry or access point handover. Different authentication mechanisms can be flexibly deployed using these standardized extensible methods. EAP methods support either terminal only or mutual authentication.

The EAP packet format is shown in Figure 4.6. The fields have the following definitions:

- Message Code – the message type code, see Table 4.6
- Se.; uer‹ ‹ # – an identifier that must match between a request and response
- Packet Length – length of the entire packet, in bytes, including all fields
- Method Code – the type code for the method extension
- Method Contents – the body of the method extension

When the message Code is EAP Success or EAP Failure, the Packet Length field is 4 and the Method Code and Method Contents fields are absent.

The EAP standard contains a few method definitions. Some of these are older, and are now out of date. They are methods from the original use of EAP on wired dialup networks that are no longer used in wireless network access control due to security threats. There are two, however, that are still important:

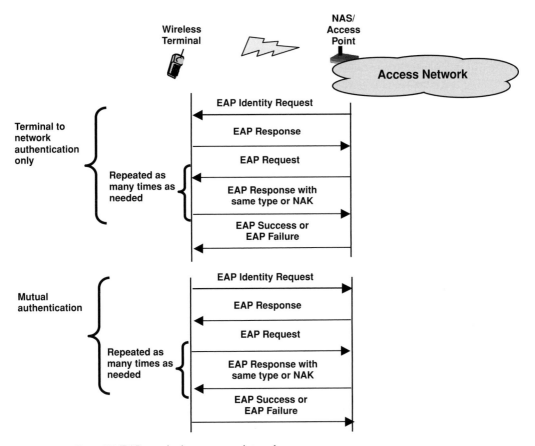

Figure 4.7 EAP terminal to access point exchange

- The Identity method type (code 0x01) is sent in an EAP Request message from the access point to the terminal for Authenticator-initiated network access control. The terminal responds with an EAP Response containing the EAP method for network access control, which the access point then forwards to the AAA server. The Identity can also be sent from the terminal to the access point if mutual authentication is desired.
- The NAK method type (code 0x03) is used when the terminal is negotiating with the network about which EAP method to use. The NAK method indicates that the EAP method included in the response is not supported.

Figure 4.7 contains the EAP message flow between the terminal and access point. After the initial EAP Identity Request, the NAS/access point forwards all EAP messages between the AAA server and terminal, unless the NAS/access point is acting as the AAA server itself. If mutual authentication is required, the authentication exchange is initiated by the terminal. The authentication exchange concludes with EAP Success or EAP Failure.

Many EAP methods have been proposed. Table 4.7 contains a list of a few that are popular. EAP-MD5 is primarily used in older, dial up systems and is not recommended for wireless systems. With the exception of PEAP, the other methods are subject to man-in-the-middle attacks and session hijacking because EAP does not support data origin authentication binding the EAP packet to the underlying transport. PEAP uses Transport Layer Security (TLS, see RFC 4346) to establish a secure tunnel with the AAA server, over which the EAP session is conducted. This eliminates the threat of session hijacking.

If authentication of the terminal at the AAA server is successful, the AAA server calculates the session master key, called the Pairwise Master Key (PMK), using a secret shared with the terminal. The terminal also calculates this key independently. In the 802.1x architecture, the AAA server provisions the PMK to the Authenticator using a Radius attribute that is encrypted using a long-term secret shared with the Authenticator, when the AAA server responds to the terminal's EAP method. If the terminal is a roaming terminal, however, the access point routes traffic through a local AAA server rather than directly to the home AAA server. The local AAA server and the home AAA server, and the local AAA server and the access point share secrets. The PMK is first sent to the local AAA server where it is decrypted before being re-encrypted and provisioned to the access point. This method is more scalable, since it does not require the access network to provision all its access points with keys shared with all its business partners, only the local AAA server needs to be provisioned with the keys. It does, however, increase the cryptographic boundary for the PMK to include the home AAA server, local AAA server, access point, and terminal. But since there are usually fewer local AAA servers than access points, the cryptographic boundaries for the access network to home network keys are reduced since fewer network entities know the access network to home network keys.

After the authentication is complete, the terminal and access point need to complete key provisioning. The 802.1x controlled port is not unblocked until this confirmation is made. This negotiation is done using a four way handshake using the EAPOL-KEY message. This negotiation performs the following functions:

- confirm that both sides are live and hold a current PMK
- derive a fresh Pairwise Temporal Key (PTK) as the root key for deriving keys for session operations
- install the session keys for encryption and data origin authentication into the 802.11 link layer
- confirm cryptographic algorithm selection
- confirm installation of the PTK.

The four way handshake is shown in Figure 4.8. The ANonce and SNonce are random or pseudorandom numbers exchanged between both sides to ensure key freshness. Note the PTK never leaves the wireless terminal and access point. The protocol also includes the provisioning of a group key for broadcast communication, but that has been dropped from the figure for simplicity.

Table 4.7 EAP methods and their characteristics

	EAP-MD5 (RFC 3748, 2004)	EAP-AKA/SIM (RFC 4187, 2006)	EAP-TLS (RFC 3748, 2004)	PEAP/EAP-TTLS
Mutual authentication	No, client authentication only	Yes	Yes	Yes, using secure tunnel
Re-keying	No, static keys only	Yes, generated during authentication and re-authentication	Yes, generated during authentication and re-authentication	Yes, generated during authentication and re-authentication
Security level	Open to offline dictionary attack	Possible man-in-the-middle hijacking in WLAN access network	Possible man-in-the-middle hijacking at WLAN access network	Protected by TLS channel (tunneling)
Ease of implementation	Simple but not for wireless	Complex, requires SIM access	PKI required	PKI required plus tunneling protocol
Applicability	Wide but not recommended due to security problems	Only for GSM/UMTS authentication servers	PKI required	PKI required plus others
Credential flexibility	Password only	USIM/SIM GSM/UMTS credentials only	Only digital certificates	Digital certificates and others

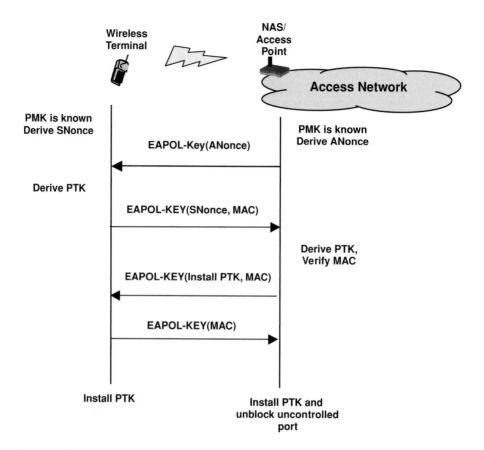

Figure 4.8 Four way handshake for completion of key provisioning

Figure 4.9 contains the format of the EAPOL-KEY Key Descriptor for 802.11. The 802.11 standard includes a number of possible key provisioning algorithms; we only consider here the three party key provisioning algorithm involving the terminal, home AAA server, and access point described above.

The fields have the following definitions:

- Key Information – a bit vector containing various bits indicating what to do with the rest of the body. Of interest here is the Install bit (bit 6) which indicates that the PTK should be installed.
- Key Length – length of the PTK in bytes. The CCMP algorithm provides data origin authentication, integrity, and confidentiality protection for 802.11 traffic using the keys established through the three party key provisioning protocol between the AAA server, the NAS, and the wireless terminal. CCMP is based on the AES encryption/decryption algorithm. The length is 16.
- Key Replay Counter – sequence number used to detect replay attacks.
- Key Nonce – a 32-byte field used to transport the ANonce and SNonce.

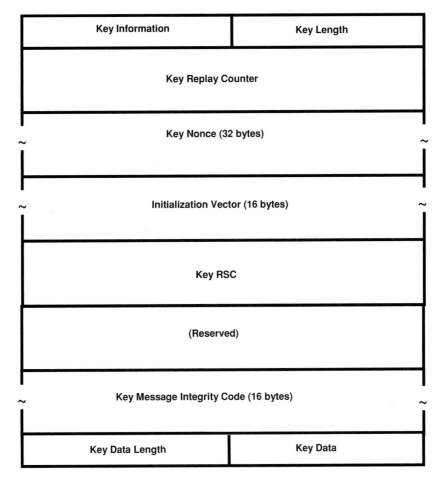

Figure 4.9 EAPOL-Key Descriptor for 802.11

- Initialization Vector – a 16 byte field used to transport an initialization vector if the key requires one, otherwise set to zero.
- Key RSC – Received Sequence Counter used with the group key. See the 802.11–2007 specification for more information (802.11, 2007).
- Key MAC – Message Authentication Code covering the entire EAPOL-Key frame from the Version field through the end of the Key Data field, with the Key Message Integrity Code field set to zero. If Version 1 of EAPOL is used, the MAC is calculated using HMAC-MD5. If Version 2 is used, HMAC-SHA1-128 is used.
- Key Data Length – length of the Key Data field, in bytes.
- Key Data – variable length field containing additional data. The encrypted group key is carried in this field; see the 802.11 standard for more information about the format.

4.4.3   Radius

The last interface in the subscription-based design is between the Authenticator/access point and the AAA server itself. If the terminal is local, the local AAA server provides authentication, authorization, and accounting service. If the terminal is roaming, the local AAA server routes the EAP traffic from the access point to the roaming terminal's home AAA server, which completes the network access control transaction. In either case, the access point and local AAA server require a security association to ensure that sensitive data, such as keys, is not divulged in transport. Similarly, the local AAA server and the home AAA server require a security association for the same reason.

Radius provides the mechanism for transporting user/terminal identity authentication, network access authorization, and key configuration between the AAA server and the access point and between AAA servers. Because Radius is a hop by hop protocol, pairwise security associations are required between every two entities communicating with Radius, providing data origin authentication, confidentiality protection, and anti-replay protection. The Radius protocol is structured as a request/response protocol with four message types. The bulk of the protocol information is carried as attribute/value pairs. Radius is carried over the Internet, using the UDP transport protocol and either IPv4 or IPv6.

Figure 4.10 contains the format of the Radius message header. The fields have the following definitions:

- Message Code – the type of message. Table 4.8 summarizes the Radius message types.
- Sequence # – used by the server for detecting duplicate requests.
- Authenticator – in an Access-Request message, a 16-bit random number expected to be unique over the lifetime of the shared secret between the Radius client and Radius server. In an Access-Accept, Access-Reject, and Access-Challenge message, an MD5 hash over the concatenation of the Code field, the Sequence # field, the Length field, the Request Authenticator field from the Access-Request packet, the response Attributes, and finally the shared secret.
- Attribute/Value pairs – a variable-length field containing the attribute/value pairs that constitute the body of the message.

Figure 4.11 provides a high-level illustration of a Radius transaction between the access point/NAS and the local AAA server. The same transaction is performed between the local AAA server and the home AAA server if the terminal is roaming.

Two types of Radius attributes are of particular interest for EAP authentication:

- EAP-Message (Attribute 79) – this attribute is used to encapsulate the end-to-end EAP protocol traffic between the terminal and the AAA server.
- The vendor-specific attributes MS-MPPE-Send-Key and MS-MPPE-Recv-Key are used for provisioning the access point with the PMK during EAP key negotiation.

Further information Radius use with 802.11 can be found in (802.11, 2007).

Table 4.8 Radius message codes

Name	Value	Description
Access-Request	0x01	Sent from the access point to the Radius server along with the terminal's EAP attribute to request network access authentication
Access-Accept	0x02	Sent from the Radius server to the access point to confirm the request for network access, along with other information needed to configure network access, such as master keys
Access-Reject	0x03	Sent from the Radius server to the access point to reject the request for network access, along with optional attributes that may be useful in the rejection
Access-Challenge	0x0b	Sent by the Radius server to the access point to challenge the terminal, for example, to test for terminals that have disappeared, and may include some attributes including relevant information. Also utilized for multiple round trips, if the Radius server requires additional information

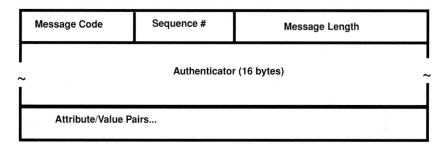

Figure 4.10 Radius message format

4.5   Hotspot design

In hotspot systems users and terminals require no prior service contract with the access network provider or a home network provider. No special AAA software is required on the terminal. Any terminal that supports a wireless link interface card and has a Web browser can obtain link access without an authentication step at the link layer. The network operator does not authenticate the user prior to network access and the only authorization necessary is authorization from a credit card company for a credit card transaction by which the access network operator can get compensation for the cost of access. No special network access control protocols are used in this design. The protocols utilized are basic Web security protocols, used for accessing secure web sites and for performing e-commerce transactions.

    The advantage of this system from the access network provider's standpoint is that it allows any Web-capable device to access the Internet, thereby increasing the potential

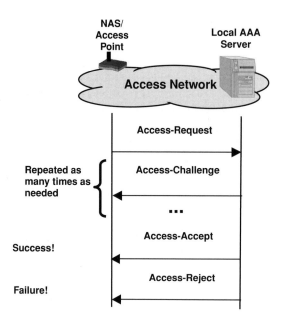

Figure 4.11 Radius exchange between NAS/access point and AAA server

customer base. Casual users, travelers, and others can obtain Internet service. This type of access control is often used in hotels, convention centers, shopping malls, and other places where people move through, and coverage is mostly localized to the particular venue. Continuity of service between the venue and surrounding areas is usually not provided. In addition, the user must interact with a Web page and a Web browser to gain network access, so this approach will not work with voice-only devices that provide Internet telephone service.

Figure 4.12 contains an overview diagram of the hotspot network system, indicating how the hotspot system implements the network interfaces from the architectural analysis in the above sections. In contrast to the subscription-based design, the access point is not involved in network access control. The Public Access Control (PAC) server disables Internet access until the user has completed a login procedure. To get to the Internet, the user must provide credentials, such as a credit card number, to a Web page allowing electronic billing of the access charge. After the credit card is verified, routing to and from the Internet is enabled and the access charge is billed to the credit card for the amount of time the terminal remains connected. The terminal to PAC Web transactions are conducted over a Transport Layer Security (TLS) secured HTTPS connection. The back end PAC to credit card server transactions are conducted over an e-commerce protocol. Because this network access control design works with any terminal supporting Web browsing, it is known as the Universal Access Method (UAM).

An important security consideration not shared with the AAA server design is that, without any other action on the part of the terminal, there is no data origin authentication or confidentiality protection for user data traffic packets over the wireless link after the HTTPS session for network access is over. In principle, this could be a serious problem

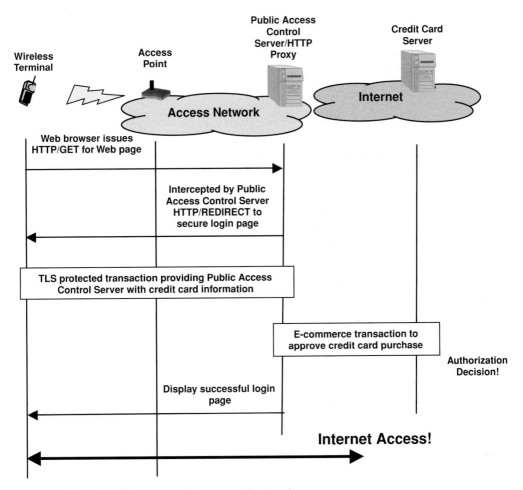

Figure 4.12 Hotspot network system and protocols

because, as mentioned in the discussion of threats above, wireless links are particularly susceptible to attacks on network access due to the open nature of wireless. However, in practice, the problem is somewhat less severe because many users with sensitive data, particularly corporate users, immediately set up a Virtual Private Network (VPN) connection between the terminal and a VPN server in a home network somewhere on the Internet. There are a variety of ways that a VPN can be established. At the IP layer, Internet Key Exchange (IKE) and IP Security (IPsec) are very common. These protocols are covered in Chapter 6, where the security architecture of IP mobility is discussed.

The hotspot design is also subject to a subtle man-in-the-middle attack. The attacker sets up a rogue access point and obtains a certificate from a certification authority whose root certificate is likely to be in a terminal's cache. When the terminal connects with the access point, the rogue sets up a TLS session and serves up a login page that looks superficially like the page for the hotspot operator. When the user types in a credit card number, the attacker steals the number or other credentials, while presenting the user with some appearance of technical problems in establishing Internet access. This attack is difficult to defend against in the hotspot design, since any laptop can advertise itself as an access point. The hotspot operator can give the user the name of the SSID for the network offline and insist that the user actually specify the SSID rather than select any SSID that comes up, but even the SSID is subject to spoofing. Other measures require more work on the part of the user or the hotspot operator.

This attack is possible due to the lack of direct authentication between the network and the user's laptop. The user's laptop does not check whether the name on the certificate matches that of the hotspot operator, it just checks whether the certificate has been authenticated from a trusted root certification authority. Certificates are easy to obtain, but spoofing the name of the hotspot operator when obtaining a certificate may be considerably more difficult. Another problem is that there is no authentication on the Web page itself. The Web page is not authenticated as coming from the hotspot operator. While it is possible for a user to check the name on a certificate after the TLS session is up but before typing in the sensitive credit card number, the appearance of a normal, secure TLS session being established probably will not be enough to trigger the user's suspicions. Clearly, more needs to be done to improve the hotspot design to resolve this problem. Requiring the hotspot operator to identify itself to the user during establishment of the TLS session would eliminate the attack, but there is no easy way to do this without changes on the host side, or, alternatively, some kind of standardized login page secured with a digital signature.

4.5.1   The TLS protocol

Unlike the subscription-based design, the hotspot design has no protocols dedicated to network access control. On the back end, there are a collection of e-commerce protocols that are used for communication between the PAC and credit card server, but these are not unique for network access control. On the front end, the terminal uses HTTP, the standard Web access protocol, to display login pages and to convey credit card information to the PAC. The only real security protocol involved in UAM access is TLS. TLS is used to secure the terminal to PAC connection so that the credit card information is protected from eavesdropping.

TLS itself is also not specifically dedicated to network access control, but it is used in the subscription-based design. Several EAP methods (EAP-TLS, PEAP, etc.) use TLS for end-to-end security between the terminal and home AAA server. In the hotspot design, TLS is used to secure the TCP transport connection. For the EAP methods, the transport connection is run over EAP. TLS has also been applied to other transport protocols.

TLS consists of two subprotocols:

- the TLS Record Protocol
- the TLS Handshake Protocol

The TLS Record Protocol is used to transport data over a transport layer, such as TCP or EAP. The TLS Record Protocol has two basic functions:

- Provide confidentiality for protocols that run over it. The security association for the TLS Record Protocol is negotiated through the TLS Handshake protocol.
- Provide data integrity through use of a keyed MAC. Cryptographic hash functions are used for generating the MAC. HMAC is used for this purpose, with either MD5 or SHA-1 to generate the message digest.

For hotspot authentication, the TLS Handshake Protocol (including the change cipher spec protocol) and the application protocol, which is HTTP, run over the TLS Record Protocol. The TLS Record Protocol specifications for data types and protocol elements are fairly complex, RFC 4346 (RFC 4346, 2006) describes the details.

The TLS Handshake Protocol is designed to negotiate a security association for the TLS Record Protocol. For hotspot authentication, the TLS Handshake Protocol negotiates security association between the terminal and the PAC for use in transmitting HTTP securely over the channel. The security association is unidirectionally authenticated, from the server to the client. The client is not required to authenticate with the server. Figure 4.13 illustrates the basic TLS Handshake Protocol as it is used in UAM. The full TLS Handshake Protocol contains more flexibility to allow, for example, client authentication. The following subsections briefly describe the TLS Handshake Protocol messages and parameters as used in UAM hotspot authentication.

ClientHello

The ClientHello message is sent from the terminal to the PAC when the terminal initiates an HTTPS session. After sending the ClientHello, the terminal waits for the ServerHello message; any other message causes termination of the protocol. The ClientHello contains the following parameters:

- The version number of the TLS protocol the client wants to use.
- A pseudorandomly generated formatted data structure.
- A session ID, which is typically empty because the client is initiating a new session with the PAC.
- A list of cryptographic algorithms supported by the client, ordered by preference. Each list entry includes a key exchange algorithm, a bulk encryption algorithm, and a MAC algorithm.
- A list of compression algorithms supported by the client, ordered by preference. This list can be empty.

ServerHello

The PAC sends a ServerHello in response to the ClientHello when it was able to find an adequate set of algorithms to set up a security association. If it cannot, it responds with a handshake error alert. The ServerHello has the following parameters:

- The version of the TLS protocol to be used. The server selects the lower version of that suggested by the terminal and the highest supported by the server.
- A pseudorandomly generated formatted data structure, separate from that provided by the terminal.

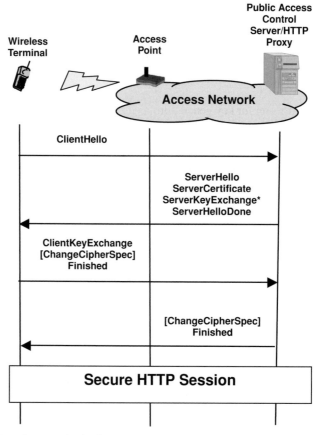

Figure 4.13 TLS Handshake Protocol as used in UAM hotspot authentication

- A session ID for identifying the session, or zero if the PAC does not want to allow resuming the session.
- A single cryptographic algorithm indicator that will be used for the security association.
- A single compression algorithm selected from the client's supplied list.

ServerCertificate

The ServerCertificate message is sent by the PAC immediately after the ServerHello. It contains a chain of certificates starting with the PAC's certificate and ending with the root certification authority. The certificates must contain a public key matching the method requested by the client, and the signing algorithm must be the same as the signing algorithm for the certificate key. Supported key exchange algorithms are RSA, Diffie–Hellman, and DSS. The supported certificate format is X.509v3.

ServerKeyExchange

The ServerKeyExchange message is sent by the PAC immediately after the ServerCertificate, if the ServerCertificate does not contain enough information to proceed with key exchange. In particular, if RSA or Diffie–Hellman key exchange is used but the key or parameters are not in the certificate, this message sends the RSA public key with which to encrypt the shared secret or the Diffie–Hellman parameters to use in generating the shared secret for key generation.

ServerHelloDone

The ServerHelloDone message is sent when the PAC has completed sending the security association parameters to the terminal. After sending this message, the PAC waits for the terminal to respond. When the terminal receives this message, it checks the server certificates and validates the other server parameters.

ClientKeyExchange

The ClientKeyExchange message is sent after the terminal has verified the PAC's parameters and completed key exchange processing. This message includes either a master secret encrypted with the RSA key or the client's Diffie–Hellman parameters for calculating a shared secret used for key generation.

ChangeCipherSpec

The ChangeCipherSpec protocol is defined in the TLS specification as a separate protocol over the TLS Record Protocol. This message indicates that the security association is about to be changed to the new association, and consists of a single byte of value 1. If a security association is in place, the old security association is used to protect the message. Typically, when the terminal and PAC are starting a session, there is no existing security association.

Finished

The Finished message is sent immediately after the ChangeCipherSpec protocol to indicate that establishment of the security association is complete. This message is the first sent protected by the security association. After it has been verified, the HTTP session for starting network access authentication is started.

4.6   Summary

Network access control is important for wireless Internet access systems because of the nature of the wireless medium. The open propagation of wireless signals means that secure network access control is necessary if the access network provider wants to limit access to the network to users that have authorization or can show the ability to pay for service. A functional architecture analysis for existing network access control systems results in three network entities: the Supplicant, the Authenticator, and the Account Authority, with appropriate functions for performing authentication and authorization.

Two general types of systems implement this architecture. One type is based on a backend AAA server as the Account Manager, the base station or access point as the Authenticator, and the terminal as Supplicant. These systems are used in public access networks where the network provider requires the user to have previously established a subscription. Another type is hotspot-based systems, where users typically are not required to have an account. Walk-up access is accommodated. AAA systems use dedicated AAA protocols running over the wireless link layer and over IP in the wired networks, while hotspot systems leverage protocols commonly used in e-commerce and other applications.

5 Local IP subnet configuration and address resolution security

After the wireless terminal has successfully obtained network access at the link layer, the next step is to obtain an IP address, last hop router address, and other parameters that allow the terminal to obtain routing service at the network layer. In turn, the last hop router uses address resolution to map the IP address of the wireless terminal to its link layer address so packets can be delivered from the Internet to the wireless terminal. Local IP subnet configuration and address resolution have a separate set of security issues that are independent from network access authentication. Even if a terminal is authenticated as a legitimate user and is authorized for service at the link layer by network access control, a rogue terminal can launch attacks on the local IP subnet configuration and address resolution processes of other terminals if these processes are not adequately secured.

In this chapter, we discuss the security of local IP subnet configuration and address resolution. After a short look at the impact of the Internet routing and addressing architecture on mobility and how that relates to local IP subnet configuration and address resolution, we briefly review the protocols for local IP subnet configuration and address resolution in IP networks, both for IPv4 and IPv6. We then discuss threats to the local IP subnet configuration and address resolution processes. We develop a functional architecture for IP subnet configuration and address resolution security based on the threat analysis and the existing protocols. Because IPv4 developed long before wireless access became common, the basic protocols for link configuration and address resolution in IPv4 are widely deployed and therefore difficult to change. Consequently, there is no real standardized protocol solution to counter the threats, though there are a few network management techniques used to mitigate attacks. On IPv6, however, the situation is much better. SEcure Neighbor Discovery (SEND) and cryptographically generated addresses provide tools for securing basic subnet configuration and address resolution in a very generic and easy-to-deploy fashion. After a brief review of security (and lack thereof) for DHCP, we focus on SEND for the remainder of the chapter.

5.1 Impact of the IP routing and addressing architecture on mobility

The Internet routing and addressing architecture requires the deployment of IP networks in subnets. IP network deployments use subnets because subnets reduce network administration effort, spread the routing load, and decrease the time required to resolve an IP address to a link layer address especially for Ethernet links. A subnet consists of one or more routers connected to other subnets and zero or more terminals to which the subnet routers provide forwarding service. All nodes on the subnet – routers and end terminals alike – are addressable from the global Internet by a *subnet prefix* in the IP address.[1] The address suffix, called the *interface identifier*, identifies the network interface hardware connecting an individual node to the link. In IPv4, the size of the subnet prefix varies in different network deployments and is determined by an additional parameter, the *subnet mask*. The subnet mask determines the number of bits in the IPv4 address that are used for the subnet prefix. The rest of the bits are used as the interface identifier (RFC 4632, 2006). In IPv6, the subnet prefix and interface identifier usually occupy the top 64 bits and bottom 64 bits, respectively, of the address (RFC 4291, 2006).

While some IP addresses have global forwarding scope, others are limited in the network topology over which they are forwarded. IPv6 supports a specific kind of address, the link local address, which is not forwarded beyond the local IP subnet. Link local addresses are heavily used in IPv6 for local IP subnet configuration. IPv4 also supports link local addresses, but these are not as widely used because they were added to the IPv4 standards more recently and are not part of the base protocol. IPv4 defines a class of addresses with forwarding limited to a local addressing realm, within a particular service provider's network. These addresses are not forwarded to the Internet. The scope of the addressing realm is controlled by a network address translator (NAT) (RFC 3022, 2001). These addresses have a certain fixed set of partial subnet prefixes, for example, 10.x.y.z ("net 10" addresses) (RFC 1918, 1996). All nodes need a globally routable IP address to communicate across the Internet, but if the addressing realm in IPv4 is behind a NAT, the NAT provides a translation between the locally routable address configured by the end terminal and a globally routable address. The end terminal only ever sees the locally routable address. IPv6 also allows limited forwarding for certain addresses, and also specifies a particular partial subnet prefix for identifying the addresses but without the use of NATs (RFC 4193, 2005).

Wireless access points typically provide link layer service only within a cell that covers a limited geographical area, so IP network service is spread over a broader geographical area by aggregating multiple access points together into a single wireless subnet. One or several last hop routers, or *access routers*, serve the subnet. Wireless terminals that move between access points on the same subnet do not change their IP subnet configuration. Wireless terminals that move to an access point in a new subnet need to perform local IP subnet configuration, a process called *IP handover* or *subnet handover*. Local subnet configuration involves obtaining an access router for routing service to the Internet and configuring one or more unicast IP addresses for the terminal's network interface card. The terminal may also configure other parameters, such as the address of a DNS server. The local subnet configuration process occurs whenever the wireless terminal enters the network for the first time and is also necessary every time the wireless terminal switches to a new access router in a new subnet. Some of the parameters, such as the DNS server,

---

[1] Routers on IPv6 networks may not be directly addressable by unicast, globally routable IPv6 addresses but they still advertise a globally routable IPv6 subnet prefix.

may remain the same if the wireless terminal does not switch to a new wireless access network provider.

The following protocols allow wireless terminals to perform local subnet configuration when they move to a new subnet:

- The Dynamic Host Configuration Protocol (DHCP) is used to query a server for local subnet configuration information. Versions of DHCP exist for both IPv4 (RFC 2131, 1997) and IPv6 (RFC 3315, 2003). DHCP provides terminals with the link layer address of an access router that forwards off the subnet, a globally routable IP address having a correct subnet prefix, and, in IPv4, a subnet mask. DHCP can also provide many other kinds of configuration information, such as the address of a DNS server.
- If the link is a serial link, the Point to Point Protocol (PPP) is used to communicate between the access router and terminal. Some wireless cellular networks also use PPP even though the connection is not over a serial link. If PPP is used, the address configuration is typically performed with the IP Configuration Protocol (IPCP) (RFC 1332, 1992), which is a part of PPP. Because PPP is often used in ways specific to a particular wireless link protocol, we do not discuss PPP further.
- IPv6 provides a protocol for local IP subnet configuration called Neighbor Discovery (RFC 4861, 2007). Neighbor Discovery supports access router information solicitation and stateless address autoconfiguration (RFC 4862, 2007). It allows an IPv6 node to autonomously generate link local and globally routable IPv6 addresses and to query for and obtain information on available access routers.

These protocols are triggered when the wireless terminal establishes a link layer connection with the network, capable of supporting IP packet communication. This connection may be a first connection after the terminal boots up, or it may be a result of a handover to a new wireless access point in a new subnet. The next section briefly reviews DHCP and Neighbor Discovery for local subnet configuration.

After the wireless terminal is configured with an IP address, packet delivery from the access router depends on the router knowing the mapping between the terminal's IP address and its link layer address. The border router and upstream routers forward packets from the Internet to the access router based on the subnet prefix. Once at the access router, however, delivery to the final destination depends on the address resolution process, mapping the interface identifier to a particular link layer address.

IPv4 and IPv6 provide different standardized protocols for address resolution. IPv4 uses a link layer protocol (standardized nevertheless as an Internet standard) called Address Resolution Protocol (ARP) (RFC 826, 1982). IPv6 uses a variation on Neighbor Discovery, the same protocol used for local IP subnet configuration. Some wireless technologies, particularly the cellular technologies, use proprietary methods for routing packets over a last hop wireless link. These methods work more like a serial link, but vary depending on the wireless technology. Since they are not standardized across different wireless technologies, they are not very general making it difficult to talk about how they fit into the Internet architecture. The next section discusses ARP and Neighbor Discovery for address resolution.

5.2   Review of local IP subnet configuration and address resolution protocols

IPv4 was originally deployed without any local IP subnet configuration protocol. Because only a small, fixed number of terminals were connected to the Internet, a terminal's IP address and access router were configured by hand. As the number of terminals grew, and especially with the introduction of wireless terminals, a server-based protocol, DHCP, was developed to handle configuration of IP addresses and other local subnet parameters. Address resolution in IPv4 uses ARP. Though this protocol deals with IP addresses, ARP is actually a link layer protocol. ARP was standardized in the early days of the Internet's development, and consequently has changed little since. Because it is so widely deployed, changes to ARP would be hard to propagate.

IPv6 was originally designed with both local IP subnet configuration and address resolution handled by Neighbor Discovery, to promote better scalability. Neighbor Discovery allows terminals and routers to autoconfigure their IP address without any human intervention. In particular, Neighbor Discovery does not require a server, reducing the administrative overhead of server maintenance. However, since many network administrators were familiar with DHCP and wanted to maintain control over IP address configuration, local IP subnet configuration with DHCP was added to IPv6. DHCP can also be used in IPv6 to configure other parameters such as the name of a DNS server. In IPv6, configuration of the IP address with Neighbor Discovery is called *stateless autoconfiguration* while configuration using DHCP is called *stateful configuration*.

5.2.1   Address Resolution Protocol

ARP was defined in the early days of the Internet in RFC 826 (RFC 826, 1982). In those days, it was an unsolved problem about how to deliver an IP packet that had been routed across the Internet to the end terminal on the last hop. When the packet arrived at the access router, the router had the IP address of the destination terminal but it needed the Ethernet address to deliver the packet. ARP allows the access router to obtain a mapping between the IP address and the Ethernet address of the network interface card that has been configured with the IP address. The access router caches the mapping in the ARP cache for some period of time, allowing the access router to perform address resolution for further incoming packets without having to perform ARP. After the cache times out, however, the access router must perform ARP again to confirm the mapping. ARP can also be used by terminals on the last hop to deliver packets directly to another terminal, rather than having to go through the access router. Figure 5.1 illustrates the protocol.

The ARP protocol runs at the Ethernet level rather than at the IP level. The access router broadcasts an ARP Request using Ethernet broadcast. The broadcast delivers the Ethernet frame to every terminal on the local link. Receiving terminals check their IP addresses, and the terminal owning the address unicasts back to the access router an ARP Reply establishing the IP address to Ethernet address mapping. The protocol is simple and direct, but as we will see in the next section, it is subject to certain threats. Note that ARP requires a link layer protocol support that broadcast. Some link layer

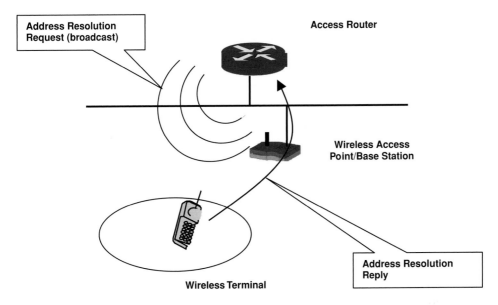

Figure 5.1 Address Resolution Protocol for IPv4

protocols (like serial links, Asynchronous Transfer Mode (ATM), and certain cellular links) do not have the capability to broadcast, and other address resolution techniques are used for these protocols.

5.2.2 Dynamic Host Configuration Protocol

DHCP for IPv4 is defined in RFC 2131 (RFC 2131, 1997) and for IPv6 in RFC 3315 (RFC 3315, 2003). Figure 5.2 illustrates the DHCP structure for a basic configuration. There are a few refinements on the basic structure, but they are by and large not relevant to local IP subnet configuration on the wireless link, with one exception discussed below. Although the basic structure of DHCP is the same in both IPv4 and IPv6, the names of the messages are slightly different. The sequence of messages is numbered in the figure, and both the IPv4 and IPv6 names are provided.

The steps involved in DHCP are (numbers keyed to Figure 5.2):

1. When a terminal arrives on a new link, the first step is to find a DHCP server. The terminal broadcasts (in IPv4) or multicasts (in IPv6) a message requesting responses from any servers that serve the link.
2. Servers respond with a message containing their IP address.
3. The terminal then selects one of the servers and sends a message describing the desired link configuration parameters. In IPv4, the terminal usually asks for one or more globally routable IP addresses, the subnet mask, and the addresses of the access routers. In IPv6, the terminal obtains access router addresses prior to DHCP using the Neighbor Discovery protocol. The terminal can also request other configuration parameters; the most essential are the addresses of DNS servers.

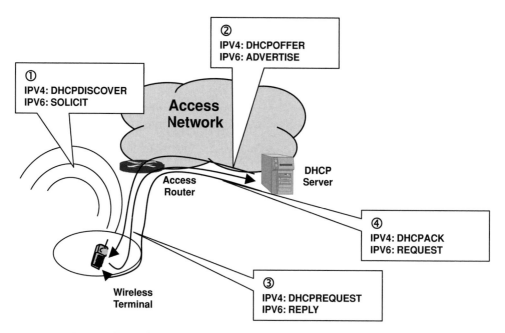

Access
Network

② IPV4: DHCPOFFER
IPV6: ADVERTISE

① IPV4: DHCPDISCOVER
IPV6: SOLICIT

Access
Router

DHCP
Server

④ IPV4: DHCPACK
IPV6: REQUEST

③ IPV4: DHCPREQUEST
IPV6: REPLY

Wireless
Terminal

Figure 5.2 Dynamic Host Configuration Protocol architecture

4. The server responds to the message requesting configuration with a message containing the requested configuration parameters, or an indication that it cannot provide them.

Not shown in the figure but important for assuring address uniqueness is duplicate detection. In order to prevent duplicate addresses, the terminal must first check whether any other terminals on the link are claiming the addresses provided by the DHCP server. This step is not required in IPv4 though it is recommended. In IPv6, however, it is required. An IPv4 terminal uses ARP to check for another holder of address. If no response is received for ARP, then the terminal knows that the address is unique on the link. An IPv6 terminal uses Duplicate Address Detection (DAD), which is part of Neighbor Discovery and is described in the next section. If a duplicate address is detected, the terminal sends a DHCPDECLINE (in IPv4) or DECLINE (in IPv6) message to the server and the server sends a DHCPACK (in IPv4) or REPLY (in IPv6) with new addresses. Address duplication occurs rarely and is usually an indication of an improperly configured server, a misbehaving terminal, or a deliberate attack.

DHCP also supports a few functions not directly associated with initial link configuration. One that is important for wireless is the confirm function, which allows the terminal to confirm whether an address previously obtained from the DHCP server is still valid. The DHCPREQUEST message performs this function in IPv4, while DHCP for IPv6 has a special message, CONFIRM, for the same function. The confirm function is particularly useful when the terminal hands over from one wireless access point to another. If the wireless access point is in the same subnet, then a quick confirmation allows the client to continue using its local subnet configuration. If not, then the client can re-run the DHCP configuration protocol to obtain a new local subnet configuration.

5.2.3   Neighbor Discovery and address autoconfiguration

In IPv6, Neighbor Discovery is the primary protocol for local IP subnet configuration. There are three basic functions involved in Neighbor Discovery:

- discovering access routers and subnet configuration information like subnet prefixes on the link;
- resolving an IP address to a link layer address for last hop packet delivery;
- autoconfiguring an IP address and checking whether any other node on the link has already claimed that address.

Neighbor Discovery uses the IPv6 version of Internet Control Message Protocol (ICMP) for the transport layer. The protocols for the Neighbor Discovery functions are discussed in the following sections.

Router discovery

When a wireless IPv6 terminal initially connects to a link, the first step in local link subnet configuration is to find an access router. The access router provides the terminal with subnet configuration information about the link. This information includes:

- The link local IP address and link layer address of the router. This allows the terminal to route traffic off of the local link and to the Internet.
- An indication of whether the terminal should use DHCP or address autoconfiguration to obtain its IP address, and whether other subnet configuration information such as the address of a DNS server is available through DHCP.
- A set of subnet prefixes that can be used to autoconfigure IP addresses, as described below.
- A collection of other information useful for managing the default router selection and address configuration.

Router discovery is a request/response protocol that provides a set of two messages: Router Solicitation (RS) and Router Advertisement (RA). The RS is multicast by the terminal to a multicast address monitored by all access routers on the subnet. The RS requests information on the access routers. The RA is a unicast reply containing that information. Note that these messages are also available in IPv4 but are typically not used strictly for router discovery.

In addition to the request/response protocol, Neighbor Discovery also supports a beacon protocol in which the access routers on a link periodically multicast an RA beacon to a dedicated multicast address. An arriving terminal listens to the multicast address and waits until the periodic RA beacons are received. The terminal collects the RAs and selects one as its default router. This process has a significant drawback for wireless terminals, in that local IP subnet configuration for IP handover may be delayed for some period while the terminal waits for an RA beacon. During that time, access to the Internet for ongoing message flows is unavailable. Minimizing the time between beacons can reduce the amount of Internet downtime, but the size of the inter-beacon period must be selected carefully so that RAs do not consume an inordinate amount of wireless bandwidth.

Resolving a link layer address to an IP address

IPv6 routers have exactly the same problem as IPv4 routers when receiving IP packets for delivery on the link: the IP address of a terminal on the link must be resolved to a local link layer address. The IPv6 routers solve the problem in roughly the same way as IPv4 routers do: the router sends out a query to all terminals on the link requesting a mapping, and the terminal owning the IP address responds with its link address. The details, however, are quite different.

Instead of using a link layer protocol, the Neighbor Discovery protocol specifies a set of IP layer messages: Neighbor Solicitation (NS) and Neighbor Advertisement (NA). The NS message is multicast out by the router to a multicast address to which all terminals on the link are listening. The NS message contains the IPv6 address of the packet which the router would like to deliver. The receiving terminal checks the IP address configured on the receiving network interface, and replies to the router with an NA containing the IP address to link address mapping if the receiving network interface is configured with the queried address. Upon receiving the NA, the router can deliver the packet. As in IPv4, terminals can also use address resolution between themselves to determine if another terminal is on link, and thereby avoid having to route packets through the access router.

Address autoconfiguration and duplicate address detection

Neighbor Discovery utilizes the router discovery and address resolution mechanism to allow autoconfiguration of addresses. Each terminal must autoconfigure at least one address: a link local IPv6 address that is only routed up to the access router, i.e. only among terminals and access routers in the local IP subnet. Link local addresses are not distributed using DHCP. Depending on the access network deployment, terminals may also autoconfigure globally routable IP addresses for access to the Internet, or they may use DHCP to obtain globally routable addresses. Which procedure to use is indicated in the RAs a terminal receives from the access routers when the terminal first arrives on the local link. The advantage of autoconfiguration is that it removes the cost and effort of managing a DHCP server for address configuration, though DHCP may be required for obtaining other information, such as the DNS configuration.

Figure 5.3 illustrates router solicitation, duplicate address detection, and address autoconfiguration for global IPv6 addresses.[2] The process has the following steps (keyed to the numbers in the figure):

---

[2] The process is similar for link local addresses, except no subnet prefix is necessary because all link local addresses have fixed prefixes.

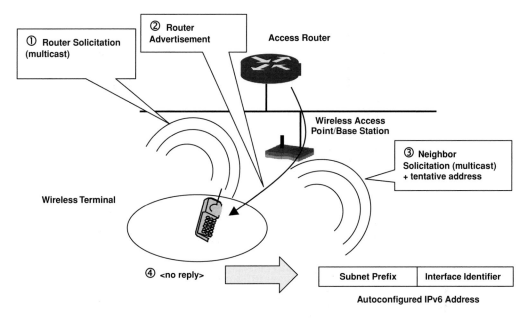

Figure 5.3 Address autoconfiguration and duplicate address detection

1. The terminal multicasts a Router Solicitation to a link local multicast address which all the access routers on the link monitor (All Routers' Multicast Address).
2. The access routers respond with Router Advertisements unicast to the terminal's link local address. The terminal selects one as its default access router.
3. The terminal constructs a tentative global unicast IPv6 address using a subnet prefix from the default router's Router Advertisement and an interface identifier. The terminal multicasts a Neighbor Solicitation with the tentative address and listens for a reply in case any other node on the link has claimed the address.
4. If no terminal responds with a Neighbor Reply, the terminal can assume the address is unique on the link, since active nodes are required to respond to Neighbor Solicitation if they own the address and nodes which are offline relinquish their right to an address. The terminal autoconfigures the network interface with the new IPv6 address and begins sending and receiving traffic.

The terminal takes the first 64 bits of the autoconfigured address from the subnet prefix advertised by the default access router, while the interface identifier part in the second 64 bits can be constructed in a variety of ways. One way recommended in RFC 4862 is to use the link layer address of the network interface card (the EUI-64 address) (Wikipedia, 2008c). Later in the chapter, we will show another way to construct the interface identifier that provides more security. Note that the duplicate address detection procedure is also required of terminals that obtain their addresses from DHCP, since the possibility of duplicate address assignment through misconfiguration also exists for statefully configured addresses.

5.2.4   IP subnet configuration and wireless link handover

Wired terminals typically do local IP subnet configuration only when they boot up. Wireless terminals, however, may need to do local IP subnet configuration for IP handover if the terminal hands over to a wireless access point in a new subnet. Some wireless link protocols and systems, particularly cellular systems, hide subnet handovers from the wireless terminal, in which case the wireless terminal's IP address does not change even if it moves to a new access router. If the wireless system requires IP handover, however, IP handover allows the wireless terminal to stay connected to the Internet as it moves, at the expense of additional work when a subnet change occurs.

When a wireless terminal is about to move out of the current wireless access point's coverage range, the terminal scans for or otherwise determines a new access point to which it can connect. In cellular networks, the terminal tells the network what access points it can "hear" and the network tells the terminal to select a particular access point. In wireless LAN networks, the terminal makes the decision which access point to select by itself. After a short handover procedure at the link layer (the exact nature of which depends on the type of wireless link protocol), the wireless terminal has connectivity to the wired network through the new access point.

If the new access point is within the same subnet as the old, the wireless terminal need perform no IP subnet configuration because the configuration performed on the old access point remains valid. The terminal must first determine whether it is, in fact, in a new subnet or not. This procedure is called *movement detection*. Movement detection in both IPv4 and IPv6 can be performed by waiting for a beaconed RA (broadcast in IPv4 and multicast in IPv6), by an RS/RA query/response, or by using DHCP. While the RA is not typically used in IPv4 for local IP subnet configuration, movement detection is performed with the RA when Mobile IP is used for IP mobility management (see next chapter). If Mobile IP is not used, then movement detection is performed with DHCP. For IPv6, movement detection is performed using either a beaconed RA or RS/RA query/response.

If subnet configuration is performed using DHCP, then a DHCPREQUEST or CONFIRM must be sent to the DHCP server to confirm that the previous address is valid. In IPv6, if address autoconfiguration is used rather than DHCP, the terminal pings the router using a RS, and determines from the returned RAs whether the subnet has changed. Alternatively, the terminal can wait for the RA beacon, at the expense of suspending its IP service until a beacon is heard. If the subnet has changed, the terminal must perform IP subnet configuration again, either using DHCP to obtain an address or using autoconfiguration.

5.2.5   Network interfaces in local IP subnet configuration and address resolution

There are three interfaces in local IP subnet configuration and address resolution:

- The router information interface between the terminal and access router over which the RA beacon and RS/RA protocol run. This interface is present in both IPv4 and IPv6 but is optional in IPv4 and only used if Mobile IP is used for IP mobility management.

- The stateful local IP subnet configuration interface between the terminal and the DHCP server.
- The address resolution interface between the terminal and any other node, including the access router.

The nature of the messaging interaction on these interfaces differs. There are three different types of messaging interaction:

- Type 1 – A broadcast/multicast solicitation message that requests a response from a DHCP server, a router, or a node that owns a particular IP address.
- Type 2 – A unicast response message from the DHCP server, the router or the node owning the queried IP address. In the case of DHCP, this message also includes a unicast request/response protocol when the terminal has discovered a server and wants to obtain additional configuration information or confirm an address.
- Type 3 – An unsolicited broadcast/multicast RA periodically providing all nodes on the link with information about the access router.

The router information interface supports both the beaconed RA protocol (interaction type 3) to provide unsolicited router and subnet information and the broadcast/multicast request and unicast response RS/RA protocol (interaction type 1 and 2) to allow the terminal to discover a router and solicit router information. The stateful local IP subnet configuration interface supports a broadcast/multicast protocol to discover a server and a request/response protocol (interaction type 1 and 2) to solicit local IP subnet configuration information. The address resolution interface consists of a broadcast/multicast protocol from a terminal requesting an address resolution followed by a unicast reply from the terminal owning the address (interaction type 1 and 2).

5.3   Threats to local IP subnet configuration and address resolution

Threat analyses have been done for both DHCP and Neighbor Discovery. No threat analysis has been done for ARP because ARP was developed before security became an important concern. At that time, the Internet consisted of a small number of fixed terminals, and Internet access was restricted to academics conducting research.

RFC 3118 (RFC 3118, 2001) briefly describes a threat analysis for DHCP as a preliminary step to defining the authentication protocol for DHCP, which we will examine later in the chapter. The RFC describes four specific threats:

- The attacker establishes a rogue DHCP server that has the intent to spoof the client with false or incorrect configuration information, for the purposes of launching a denial-of-service attack or man-in-the-middle attack.
- Related to the above is an inadvertent attack caused by a mistakenly configured server. In this case, the attack is not intentional but the practical effect on users is similar.
- An invalid client masquerades as a valid client to steal IP service or otherwise circumvent auditing.

- A denial-of-service attack in which the attacker exhausts claimable resources such as addresses by continually requesting them.

Specific mitigation measures recommended by RFC 3118 are the following:

- Network access control filters out clients that have no authorization for network access, mitigating any threat from invalid clients. In hotspot networks, which do not support network access control, this threat remains.
- All protocols experience the denial-of-service threat, and RFC 3118 recommends redundancy as the primary mitigation measure.

The residual threats to DHCP come from rogue and misconfigured DHCP servers. These threats are possible even in tightly controlled enterprise networks.

RFC 3756 (RFC 3756, 2004) provides a comprehensive analysis of threats for IPv6 Neighbor Discovery. The RFC separates threats into three different classes based on the functionality provided by the Neighbor Discovery Protocol and location:

- Threats against the basic address resolution and address autoconfiguration functions of Neighbor Discovery. These functions do not involve routers, and any attacks must be launched locally because routing of Neighbor Discovery packets is restricted to the local link.
- Threats against the router solicitation and advertisement functions of Neighbor Discovery. Attacks on these functions must be launched locally for the same reason is in the first bullet point.
- Threats involving replay attacks or attacks that can be launched remotely. In general, these are considered to be more serious since discovering and disabling the attacker is often more difficult if the attack is not confined to the local link.

We discuss each class in the following subsections.

5.3.1   Threats against address resolution and autoconfiguration

One threat to address resolution and autoconfiguration is spoofing of NS and NA messages. Nodes on the link, including the router, use NS/NA to create a binding between the IP address and link address, so packets can be delivered over the last hop. NS/NA is also used in duplicate address detection to ensure that no other node on the link has the address.

An attacking node spoofing an NA can cause packets to be delivered to another link address, where the packets can be siphoned off and processed under the control of the attacker. The attacker can also deny the possession of an IP address to a node by spoofing NAs during duplicate address detection. Since RFC 4862 says that if duplicate address detection fails after three tries a node should give up, this effectively denies the node IP service, resulting in a DoS attack.

An attack on NS/NA can also be used to thwart neighbor unreachability detection. Normally, if a node does not receive a reply to a message after 30–50 seconds (depending on configuration), it will invoke the neighbor unreachability procedure. This procedure involves sending a unicast NS to the address in question. If the node possessing the address is still reachable, it will reply with an NA. The soliciting node tries several times if no reply is immediately received, but eventually, the binding between the IP address and link address is deleted if the target node does not answer.

An attacker can disrupt neighbor unreachability detection by sending fabricated NAs in response to a neighbor unreachability detection NS message. By doing this, the victim believes that the address in question is still available when it is actually gone. This constitutes a kind of DoS attack, since the victim will uselessly continue to try to communicate rather than break off the communication and attempt connecting with an active node. The attacker can also use this attack to forage the address mapping for a third party, thereby causing the victim to be deluged with unwanted traffic.

5.3.2 Threats against router discovery and routing

The primary threat against router discovery is that a malicious node masquerades as a router. The attacker responds to RS messages from nodes on the link requesting router discovery with bogus RA messages, giving its own link layer address and link local IPv6 address as a router address. The attacker can also multicast periodic bogus RA messages, thereby spoofing nodes that are listening for the RA beacon on the link. The attacker can also cause nodes that have selected a legitimate router as the default to drop the legitimate router by multicasting RA beacons for the legitimate router with a lifetime of zero, thereby causing the victim node to select the attacker as the default router. Once a node has accepted the attacker as a default router, the attacker can manipulate the victim's traffic at its leisure. Packets can be inspected, service can be denied, etc.

Another attack involves compromising a legitimate last hop router, either by shutting the router down or by taking control of it. If the last hop router is killed, nodes on the link attempt to find another router after a short delay. The attacker can advertise itself as a router. If a trusted router is taken over by an attacker, the attacker can then examine traffic, exactly the same as if the attacker had convinced the nodes on the link to accept it as a legitimate router in the first place. These attacks are hard to protect against in system and protocol design.

Another more subtle attack involves advertising false parameters in RAs, like the wrong subnet prefix or an indication that the link requires DHCP when it really does not. A victim node that uses the false parameters for local IP subnet configuration would then be unable to obtain IP routing service, or, in the case of DHCP, a bogus DHCP server could hand out the address of a man-in-the-middle attacker or otherwise redirect traffic. This attack is similar to the bogus router attack, but does not require the attacker to actually advertise itself as a router in order to disrupt traffic.

5.3.3 Replay and remote attacks

Neighbor Discovery protocol messages have no protection against replay attacks. This lack of protection allows an attacker to record and replay out-of-date messages in order to spoof the victim. For example, an attacker can record an NA from a node, and later can modify the NA and replay it to disrupt the last hop routing for the node that originated the NA.

As with all IP protocols, DoS attacks on Neighbor Discovery are possible. A node anywhere on the Internet can fabricate addresses and bombard the router with traffic for some protocol (for example HTTP) having the bogus addresses. This will cause the router to multicast NS messages for the addresses, which never get any response because the addresses are fabricated. A terminal trying to enter the network could be rendered unable to perform duplicate address detection due to the increased traffic. This attack can be mitigated if the router rate limits NS traffic after a certain threshold has been reached.

5.3.4   Security services for countering threats

The security services necessary for countering the threats are the following:

- Identity verification and authorization is required on address resolution traffic to ensure that signaling for address resolution originated at an IP address that is authorized to claim the address.
- Identity verification and authorization is required on router advertisement to ensure that the router is authorized to route traffic.
- Data origin authentication is required on address configuration (including address autoconfiguration) and address resolution traffic to ensure that the message originated at the claimed address and that the address was not modified while in transit. Data origin authentication is also required on Router Advertisement traffic for the same reason.
- Identity management and key management is required on address configuration (including address autoconfiguration), address resolution, and Router Advertisement traffic to set up the security association for data origin authentication and to supply any material required for identity verification and authorization.

Note that the security services required for local IP subnet configuration and address resolution do not include confidentiality. Confidentiality is not required because the information conveyed – IP addresses for the wireless terminal and router, other subnet parameters, etc. – is public information and is known and discoverable to other nodes on the subnet and on the Internet. In some cases, such as duplicate address detection, the information must be known to all nodes on the subnet in order for the protocol to function correctly.

5.4   Functional architecture for local IP subnet configuration and address resolution security

The existing local IP subnet configuration and address resolution protocols dictate the following functional entities for local IP subnet configuration and address resolution:

- the Basic IP Node, which could be a wireless terminal capable of moving from one network to another, or a router or configuration server acting as a simple IP node;
- the Access Router, the last hop router for the wireless terminal;
- a Local Subnet Configuration Server, which provides configuration information to the wireless terminal.

Note that access routers and local subnet configuration servers must support certain Basic IP Node functions in addition to their own functions, since they act as basic IP nodes when communicating with other IP nodes or configuring their own local subnet information.

The sections below describe the functions associated with these functional entities.

5.4.1 Functional architecture and interfaces

Figure 5.4 illustrates the functional architecture of the local IP subnet configuration and address resolution security system. Only network interfaces are shown. Programmatic interfaces may exist between the security functions and the communication functions, credential storage and generation, or for other functions depending on implementation.

Four interfaces require protocol definitions:

- BN1 – the interface between all Basic IP Node functional entities on the local subnet. This interface is responsible for security of address resolution and, in the case of IPv6, address autoconfiguration and duplicate address detection. In IPv4 the protected messages are ARP, ARP Reply, and Router Solicitation. In IPv6 the protected messages are Neighbor Solicitation, Neighbor Advertisement, and Router Solicitation.
- AR1 – the interface between the Access Router and Basic IP Node. This interface is responsible for credential exchange from the Access Router to the Basic IP Node and for authentication of the router and local subnet configuration information message, the Router Advertisement.
- LCS1 – the interface between the Local Subnet Configuration Server and the Basic IP Node for authentication. The protected messages are the DHCPv4 or DHCPv6 messages.
- LCS2 – the interface between the Local Subnet Configuration Server and the Basic IP Node for key provisioning and identity management. This interface provides a protocol for authenticated credential exchange between the Basic IP Node and Local Subnet Configuration Server for credentials to protect the DHCP traffic.

The interfaces here are between functional entities. In an actual implementation, both the access router and DHCP server also support BN1. This is because both entities must support address resolution and, in IPv6, address autoconfiguration and duplicate address detection, just like any other IP node on the subnet. Note also that there is no interface between IP nodes specifically for credential exchange. Depending on the actual security protocol, credential exchange may be necessary, but this can be included with the actual

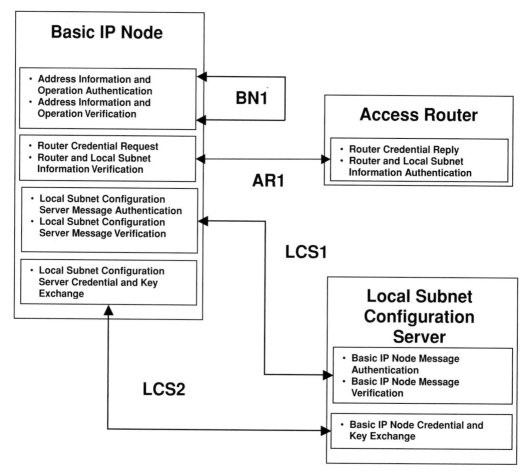

Figure 5.4 Functional architecture of local subnet configuration and address resolution security system

messaging for address resolution and address autoconfiguration, as part of the BN1 interface.

5.4.2   Basic IP Node functions

Table 5.1 contains a list of functions, security services, parameters, and return objects supported by the Basic IP Node for securing local IP subnet configuration and address resolution. The following subsections describe the functions in more detail.

Local Subnet Configuration Server Message Authentication function

The Local Subnet Configuration Server Message Authentication function takes a message to the Local Subnet Configuration Server and the credentials shared with the Local Subnet Configuration Server and returns the message with authentication information in

Table 5.1 Functions, parameters and results for Basic IP Node

Function	Security services	Parameters	Return
Local Subnet Configuration Server Message Authentication	– Data origin authentication on signaling to the Local Subnet Configuration Server	– Clear text message to the Configuration Server – Long-term credentials shared with the Local Subnet Configuration Server (e.g. configuration token, shared key, certified public key, etc.)	– Message to the Local Subnet Configuration Server including authentication information
Local Subnet Configuration Server Message Verification	– Data origin verification on signaling from the Local Subnet Configuration Server	– Local Subnet Configuration Server identity – Clear text message from the Local Subnet Configuration Server containing authentication information – Long-term credentials shared with the Local Subnet Configuration Server (e.g. configuration token, shared key, certified public key, etc.)	– Yes/no indication whether the message verification succeeded
Local Subnet Configuration Server Credential and Key Exchange	– Identity and key management for the security association with the Local Subnet Configuration Server	– Credentials needed for exchange with the Local Subnet Configuration Server	– Local Subnet Configuration Server credentials and authentication key
Access Router Credential Request	– Identity management and authorization check for the Access Router	– Access Router identity	– Credentials needed to verify Access Router
Access Router and Local Subnet Information Verification	– Data origin verification on signaling from the Access Router	– Access Router identity – Access Router credentials – Access Router and local subnet information message with authentication	– Yes/no indication whether the message verification succeeded
Address Information and Operation Authentication	– Data origin authentication on signaling operating on the Basic IP Node's address	– Message asserting information about or operating upon the node's IP address – This node's credentials proving the node's right to the address	– Message asserting information about or operating upon the IP node's IP address with authentication
Address Information and Operation Verification	– Data origin verification on signaling from another Basic IP Node claiming right to operate upon the other node's address	– Message asserting information about or operating upon another node's IP address, with authentication from the other node – The other node's credentials proving the node's right to the address	– Yes/no indication whether the message verification succeeded

it allowing the Local Subnet Configuration Server to verify that the message came from the authorized node.

Local Subnet Configuration Server Message Verification function

The Local Subnet Configuration Server Message Verification function takes a message from the Local Subnet Configuration Server containing authentication information, the Local Subnet Configuration Server identity, and the credentials shared with the Local Subnet Configuration Server. The function returns an indication of whether the message was verified as having originated with the Configuration Server.

Local Subnet Configuration Server Credential Exchange function

The Local Subnet Configuration Server Credential Exchange function takes credentials for this node to exchange with the Local Subnet Configuration Server and returns the Local Subnet Configuration Server credentials and an authentication key for authenticating exchanges with the Local Subnet Configuration Server. Depending on the implementation, other parameters may be required.

Access Router Credential Request function

The Access Router Credential Request function takes an indication of the identity of the Access Router for which credentials are desired and returns the credentials needed to verify a message from the Router.

Access Router and Local Subnet Information Verification function

The Access Router and Local Subnet Information Verification function takes the Access Router's identity, the Access Router's credentials, and a message from the router containing authentication information requiring verification, and returns an indication of whether the message verification succeeded.

Address Information and Operation Authentication function

The Address Information and Operation Authentication function takes a message containing information about or directions for operating upon the node's IP address. This information is, for example, the mapping between the node's IP address and link layer address. The function also takes the node's credentials proving its right to the address. The function returns the message with authentication proving the message originated from the node having the right to the address.

Address Information and Operation Verification function

The Address Information and Operation Verification function takes a message from another Basic IP Node asserting some information or directions for operating on the other node's IP address containing authentication, and the other node's credentials proving its right to the address. The function returns a yes/no indication of whether the message originated from the node.

Table 5.2 Functions, parameters, and results for Access Router

Function	Security services	Parameters	Return
Access Router Credential Reply	– Identity management and authorization with the Basic IP Node	– Message from the Basic IP Node requesting router credentials – Access Router credentials sufficient for an IP node to verify authorization to route	– Message to the Basic IP Node containing router credentials and including authentication
Access Router and Local Subnet Information Authentication	– Data origin authentication on router advertisement signaling	– Access Router identity – Access Router credentials – Access Router and local subnet information message in clear text	– Access Router and local subnet information message including authentication

5.4.3   Access Router functions

Table 5.2 contains functions, security services, parameters, and return objects for the Access Router. Note that Basic IP Node functions are not included in this table, though an implementation of an Access Router must support those also because routers are, themselves, additionally IP nodes. The following subsections describe the functions in more detail.

Access Router Credential Reply function

The Access Router Credential Reply function responds to a request from a Basic IP Node for router credentials. The parameters are the message from the IP node requesting Access Router credentials and the Access Router's credentials. The function returns the credential reply message including any authentication.

Access Router and Local Subnet Information Authentication function

The Access Router and Local Subnet Information Authentication function generates an authenticated router and local subnet information message either in response to a solicitation from an IP node on the local subnet or autonomously for use as a periodic beacon. The parameters are the Access Router's identity, the Access Router's credentials, and the Access Router and local subnet information message in clear text. The return is the Access Router and local subnet information message including authentication.

5.4.4   Local Subnet Configuration Server functions

Table 5.3 contains the functions, security services, parameters, and return values for the Local Subnet Configuration Server. As with the Access Router, the Basic IP Node functions are not included in the list, even though the Local Subnet Configuration Server needs them to act as a proper IP node on the local subnet. The following subsections describe the functions in more detail.

Table 5.3 Functions, parameters and results for Local Configuration Server

Function	Security services	Parameters	Return
Basic IP Node Message Authentication	– Data origin authentication on signaling to the Basic IP Node	– Clear text message to the Basic IP Node – Long-term credentials shared with the Basic IP Node (e.g. configuration token, shared key, certified public key, etc.)	– Message to the Basic IP Node including authentication information
Basic IP Node Message Verification	– Data origin verification on signaling from the Basic IP Node	– Basic IP Node identity – Clear text message from the Basic IP Node containing authentication information – Long-term credentials shared with the Basic IP Node (e.g. configuration token, shared key, certified public key, etc.)	– Yes/no indication whether the message verification succeeded
Basic IP Node Credential and Key Exchange	– Identity and key management for the security association with the Basic IP Node	– Credentials needed for exchange with the Basic IP Node	– Basic IP Node exchange credentials and authentication key

Basic IP Node Message Authentication function

The Basic IP Node Message Authentication function generates authentication on a message replying to a Basic IP Node message request. The parameters are the clear text message to the Basic IP Node and the long-term credentials shared with the Basic IP Node. The return is the message including authentication information allowing the Basic IP node to authenticate the message.

Basic IP Node Message Verification function

The Basic IP Node Message Verification function verifies a message from a Basic IP Node. The parameters are the Basic IP Node identity, a clear text message from the Basic IP Node containing authentication, the long-term credentials shared with the Basic IP Node. The return is a yes/no indication of whether the message authentication succeeded.

Basic IP Node Credential Exchange Reply function

The Basic IP Node Credential Exchange Reply function takes credentials for the Local Subnet Configuration Server to exchange with the Basic IP Node and returns the Basic IP Node credentials and an authentication key for authenticating exchanges with the Basic IP Node. The exact parameters and return values depend on the credentials used for authentication, for example, public key certificates may be exchanged or a shared key may be configured.

Table 5.4 Mapping of interfaces to protocols for IPv4 and IPv6

	Protocols on interfaces			
	BN1	AR1	LCS1	LCS2
IPv4	– < None >	– < None >	– Authentication Option for DHCP (RFC 3118)	– Manual Configuration
IPv6	– Secure Neighbor Discovery (RFC 3971)	– Secure Neighbor Discovery (RFC 3971)	– Authentication Option for DHCP (RFC 3118) – IPsec AH (RFC 4302) or ESP Authentication (RFC 4303)	– Manual Configuration – IKEv1 (RFC 2409) – IKEv2 (RFC 4306)

5.4.5 Taxonomy of deployed systems

Table 5.4 provides a mapping between the interfaces in Figure 5.4 and the protocols in IPv4 and IPv6 that implement the interfaces. Of particular note is the lack of any security protocols for IPv4 on the BN1 and AR1 interfaces. As mentioned above, the ARP protocol was developed in the early days of the Internet before security was considered an important issue, and therefore there is no security protocol for ARP. Similarly, the IPv4 Router Discovery messages specified in RFC 1256 (RFC 1256, 1991) contain no cryptographic protection though there are a few rough security rules that prevent simple attacks. We discuss these and mitigation measures for attacks on ARP below.

On the LCS1 and LCS2 interfaces, RFC 3118 describes a standardized authentication option for DHCP, but does not provide any credential or key exchange. Manual, out-of-band configuration is recommended. RFC 3118 applies to either DHCPv4 or DHCPv6; however, DHCPv6 also recommends using IKE for credential and key exchange and IPsec for data origin authentication protection. The DHCP authentication option is discussed below, IKE and IPsec are discussed in Chapter 6.

5.5 Security protocols for address resolution, address autoconfiguration, and router discovery

Security on the BN1 and AR1 interfaces is handled differently in IPv4 and IPv6. There are no formal cryptographic protocols used in IPv4 for securing these interfaces, so security is provided using operational rules. In contrast, the IPv6 Secure Neighbor Discovery protocol provides cryptographic protection against attacks on address resolution, address autoconfiguration, and router discovery. The next two sections provide details.

5.5.1 Security for address resolution and router discovery in IPv4

IPv4 address resolution and router discovery use some heuristic, operational rules to mitigate attacks on the BN1 and AR1 interfaces. IPv4 provides no support for address autoconfiguration, so no security is required.

Address resolution in IPv4 is performed by the Address Resolution Protocol (ARP). Attacks on ARP are called "ARP spoofing." In these attacks, the attacker replies to an ARP with its IP address, causing the node sending the ARP to install an incorrect mapping in the mapping table, called the ARP cache. The attacker can then insert itself as a man in the middle and inspect traffic between the two nodes. One way to prevent ARP spoofing is to not use the protocol, and instead install fixed tables in the routers that resolve the IP address to the link address. However, this technique is not practical in wireless networks, since IP addresses are assigned dynamically, and subnets can support many clients.

A more practical method that is commonly used in existing products is DHCP snooping combined with ARP inspection and validation. The switches within a switched LAN monitor DHCP traffic across untrusted ports when an IPv4 node initially configures its IP address. The switches record the valid link layer address to IP address bindings seen in the DHCP replies on the monitored ports. Later, when an ARP reply is seen on a port, the switch compares the IP address and link layer address in the reply to the recorded addresses, and if the two do not match, the ARP reply is dropped. This prevents an attacker from substituting its link layer address for the victim's. This defense fails if the attacker changes its link layer address to match the victim's. The IP to link layer address mapping on the switch matches the ARP reply, but link layer delivery of packets is disrupted because there are now two network interface cards with the same link layer address on the link. If the attacker's intention is to disrupt packets to the victim, this will certainly do it. Use of 802.1x network access authentication is the only way to deter link address spoofing, because 802.1x locks down what link addresses are allowed on specific ports.

Router discovery in IPv4 supports a set of rules to determine whether or not a Router Advertisement is valid but the rules do not protect against any attacks. They merely determine whether the message is well-formed. The security section of RFC 1256 mentions that signed RAs are an item for further study, but no additional security protocols have been added to router discovery. Mobile IPv4 does define an RA extension that provides a weak form of replay protection for registrations with the access router (called a "foreign agent" in Mobile IPv4). As a practical matter, router discovery is not used in IPv4 except by Mobile IPv4. IPv4 nodes typically obtain their access router link address and IP address from DHCP.

The security measures developed for ARP are a good illustration of what happens when no security architecture is developed for a system. The lack of a security architecture and protocols to implement the architecture means the system is vulnerable to attacks when deployed. Since it is difficult to modify a protocol after it has been widely deployed, vendors often develop nonstandard patches that provide a measure of security but in a way that is sometimes not interoperable and invariably much less straightforward than if the proper cryptographic protections and other security measures were included in the protocol in the first place. Sometimes, standards are developed to increase interoperability, but standards put in place after widespread deployment are easy to ignore and must often compromise simplicity for backward compatibility. In the case of ARP, the fact that the protocol is at the link layer, which is difficult to change, and is widely deployed in many routers means that any simple cryptographic solution is unlikely to be deployed. The combination of administrative and deployment patches mentioned above are the only substitute.

5.5.2 Security for address resolution, address autoconfiguration and router discovery in IPv6

In contrast to IPv4, the design of IPv6 address resolution and router discovery considered security from the beginning. Security for address autoconfiguration was required as well. The original IPv6 Neighbor Discovery protocol specification in RFC 2461 (RFC 2461, 1998) and the address autoconfiguration protocol in RFC 2462 (RFC 2462, 1998) require use of IPsec (RFC 4301, 2005) for security. Because Neighbor Discovery is at the IP layer, unlike ARP, IP level security can in theory be used to secure it.

However, subsequent study determined that IPsec was not a good match for Neighbor Discovery security. IPsec was developed for one-to-one security associations developed between two specific terminals. Traffic for address resolution and address autoconfiguration has more of a one-to-many nature, i.e. multicast. In addition, the IPsec security associations are usually intended to last for a longer period between terminals that are exchanging traffic frequently or at least have the potential to do so. Along with router discovery, the traffic profile of address resolution and address autoconfiguration is more ephemeral. A node performs router discovery, address autoconfiguration, and address resolution when it first comes upon a new link, but afterward, these operations are done at periodic but very infrequent intervals, purely to refresh the internal caches of IP address to link address mappings and the list of available last hop routers.

As a consequence, a new protocol for securing Neighbor Discovery was developed with characteristics more in tune with the ephemeral nature of the Neighbor Discovery traffic profile. The protocol is called SEcure Neighbor Discovery (SEND), and is documented in RFC 3971 (RFC 3971, 2005). RFC 3972 (RFC 3972, 2005) describes a new security technique called Cryptographically Generated Addresses (CGAs) which forms the basis of SEND. These topics are discussed in the next two sections. When Neighbor Discovery protocol and address autoconfiguration were updated in RFC 4861 (RFC 4861, 2007) and 4862 (RFC 4862, 2007), SEND was recommended for security, except in cases where the IP address mappings are statically configured.

Cryptographically Generated Addresses

Cryptographically Generated Addresses (CGAs) are a key component of SEND. CGAs solve the problem of binding a cryptographic signature to an IPv6 address. This binding is necessary so that the recipient of a NA can verify that the NA was sent by an IPv6 node with authorization to claim the address. The recipient can then verify the authorization of the sending node to operate upon the address in some way. For address autoconfiguration, the authorization asserts ownership of the address in response to a query from another node about wanting to use it. For address resolution, the authorization secures the IPv6 to link layer address mapping, so that no other node can spoof the last hop route to the address' rightful owner. CGAs' applicability is not just restricted to SEND. They have

Table 5.5 Values of *Mask2* for Different Values
of the *Sec* Parameter

Sec	Mask2
0	0x00000000000000000000000000000000
1	0xffff0000000000000000000000000000
2	0xffffffff0000000000000000000000000
3	0xffffffffffff000000000000000000
4	0xffffffffffffffff000000000000
5	0xffffffffffffffffffff00000000
6	0xffffffffffffffffffffffff0000
7	0xffffffffffffffffffffffffffff

been used in a variety of enhancements to Internet protocols, where operations on an IPv6 address are required with authorization from the address owner.

CGAs are based on using the hash of a public key generated by the node claiming the address, together with a few other parameters, to form the interface identifier (last 64 bits) of the IPv6 address. The public key can be either certified or generated ad hoc simply for address construction. The private key is then used to sign the signaling message that performs some operation upon the address. The address owner sending the message includes the public key and the other parameters along with the message. A receiving node verifies the message by first checking whether the message originated from a node possessing the public key. The check operation involves hashing the public key and combining it with the other parameters including the subnet prefix to form a test CGA. If the test CGA matches the address on the message, then the receiving node knows that the message originated from the node possessing the public key. The receiving node then checks the public key signature. If the signature verifies, the receiving node knows the message originated with the owner of the matching private key and that the message was not modified in transit, verifying the authorization of the sending node to claim the address.

In order to provide protection against increasing processor speed (and thus the ability of an attacker to mount a real-time birthday attack), RFC 3972 defines a standardized security parameter, called *Sec*. The *Sec* parameter is included in the address. *Sec* is a three-bit unsigned integer used in the generation of the CGA and encoded into the three leftmost bits of the interface identifier. The *Sec* parameter controls the value of the 112-bit hash mask 2 (*Mask2*) used in the CGA construction algorithm. Table 5.5 contains the values of *Mask2* for each *Sec* value.

The algorithm also uses another hash mask, *Mask1*, of 64 bits to mask off bits 0, 1, 2, 6, and 7. *Mask1* has value 0x1cffffffffffffff. Bits 0 through 2 constitute the *Sec* parameter; bits 6 and 7 are the EUI 64 "universal/local" ("u") and "individual/group" ("g") bits. EUI 64 is an IEEE standard format for Ethernet link layer addresses (Wikipedia, 2008c). Since EUI 64 identifiers are commonly used for generating nonCGAs, the CGA algorithm maintains compatibility with the "u" and "g" bits. For CGAs, the "u" and "g" bits are set to zero. The rest of the bits in the interface identifier, 59 in all, are generated by hashing the public key and other parameters as described below.

More formally, if *Hash1* and *Hash2* are hashes of the public key and other parameters constructed as described below, a CGA is an address whose interface identifier satisfies the following two conditions:

1. *Hash1* & *Mask1* $==$ interface identifier & *Mask1*
2. *Hash2* & *Mask2* $==$ 0x00000000000000000000000000000

These conditions are used by the receiving node to check the authenticity of a CGA. Generation of a CGA requires three values:

- a 64-bit IPv6 subnet prefix for the subnet in which the CGA will be topologically located;
- the generating node's public key in the format into which the generating node encodes it to send to the receiver;
- the security parameter *Sec*.

The algorithm for generating a CGA is as follows:

1. Generate a random or pseudorandom 128-bit modifier value.
2. Concatentate from left to right the following: modifier | 9 zero bytes | the encoded public key | any optional extension fields.
3. Execute the SHA-1 algorithm on the concatenation and take the leftmost 112 bits of the SHA-1 hash value. Set this to *Hash2*.
4. Compare the $16 * Sec$ leftmost bits of *Hash2* to zero. If they are all zero or *Sec* is 0, continue with Step 5; otherwise, increment the modifier by 1 and go back to Step 2.
5. Set the 8-bit collision count to zero.
6. Concatenate from left to right the following: final modifier | subnet prefix | collision count | encoded public key | any optional extension fields.
7. Execute SHA-1 on the concatenation and take the 64 leftmost bits of the SHA-1 hash value. Set this to *Hash1*.
8. Form an interface identifier from *Hash1* by writing the value of *Sec* into the three leftmost bits and by setting bits 6 & 7 (i.e. the "u" and "g" bits) to zero.
9. Concatenate the subnet prefix and interface identifier together to form the address: subnet prefix | interface identifier.
10. Perform duplicate address detection as defined in RFC 4862 if necessary. If an address collision is detected, increment the collision count by one and go back to Step 6. After three collisions have been detected, report an error (it may be a denial-of-service attack).
11. The resulting CGA is now ready for use. The parameters used to generate the CGA – the final modifier value, the subnet prefix, the final collision count, the encoded public key, and any optional extension fields – are concatenated together to form the CGA Parameters option (described below) and sent along with the signed message.

From the above algorithm, it should be clear how the *Sec* parameter helps protect against increasing processor power at the service of the attacker (so-called "Moore's

Law" protection). If *Sec* is zero, the algorithm is deterministic and relatively fast. If *Sec* is not zero, however, the algorithm is not guaranteed to terminate after a certain number of iterations, since the algorithm is effectively conducting a brute-force search of the space of modifier values for particular values that match the zero bits criteria, though the algorithm does take $O(2^{16 \times Sec})$ iterations. While the search criteria impose a performance constraint on a legitimate generator of the CGA, the search is typically only conducted once, when the address is generated. An attacker would have to conduct the search every time it was testing a potential attack address. This procedure helps protect against a birthday attack even though the number of hash bits in the interface identifier is relatively small (59) by making the attack computationally expensive.

Verification of a CGA requires the IPv6 address to be tested and a CGA Parameters option with the parameters the generating node used to construct the CGA. The verification consists of the following steps. Each step must succeed for the verification to succeed, and the verification is stopped if any step fails:

1. Check that the collision count in the CGA Parameters option is 0, 1, or 2. If any other value is found, the verification fails.
2. Check that the subnet prefix in the CGA Parameters option is the same as the subnet prefix in the address. If the two do not match, the verification fails.
3. Execute the SHA-1 algorithm on the CGA Parameters option, and take the leftmost 64 bits of the hash value. Set this to *Hash1*.
4. Mask *Hash1* with *Mask1*, mask the interface identifier with *Mask1* and compare the two values. If they do not match, the verification fails.
5. Form the *Sec* parameter as an unsigned integer from the leftmost 3 bits of the interface identifier.
6. Concatenate from left to right the following, retrieving values from the CGA Parameters option: the modifier | 9 zero bytes | the encoded public key | any extension fields in the CGA Parameters option corresponding to additional parameters.
7. Execute the SHA-1 algorithm on the concatenation, and take the leftmost 112 bits of the hash value. Set this to *Hash2*.
8. Form *Mask2* from Table 5.5 according to the value of the *Sec* parameter. Mask *Hash2* with *Mask2*. If this value is not zero, the verification fails. Note that if *Sec* = 0, the verification never fails at this step.

The CGA Parameters option is a data structure which carries the concatenated parameters between the generating node sending a message and the receiving node. Figure 5.5 illustrates the format of the CGA Parameters option. The fields have the following values:

- Modifier – a 128-bit randomly generated unsigned integer used during CGA generation.
- Subnet Prefix – the 64-bit subnet prefix of the subnet in which the CGA is topologically located.

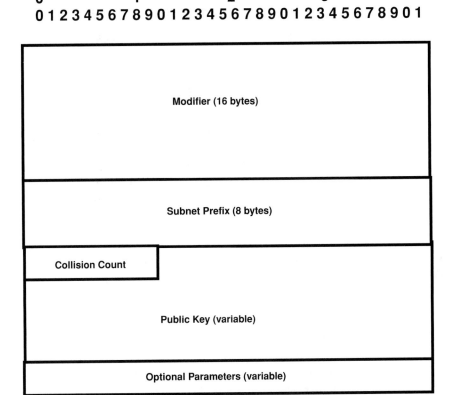

Figure 5.5 Format of the CGA Parameters option data structure

- Collision Count – an 8-bit unsigned integer that must be 0, 1, or 2. The Collision Count is incremented during CGA generation if a collision is detected during duplicate address detection.
- Public Key – a variable-length field containing the generating node's public key formatted as a DER-encoded ASN.1 structure of type SubjectPublicKeyInfo. This format is defined in the Internet X.509 certificate profile in RFC 3280.
- Optional Parameters – a variable-length field containing any optional parameters used in the CGA generation.

Full protection of a message containing a CGA requires that the digital signature be calculated and verified in addition to the CGA. RFC 3972 requires that each application of CGAs obtain a 128-bit type tag from the CGA Message Type name space, in order to differentiate signatures. For SEND, the CGA type tag is 0x 086f ca5e 10b2 00c9 9c8c e001 6427 7c08. The type tag is concatenated with the message and the signature is calculated using the RSASSA-PKCS1-v1_5 signature algorithm (defined in RFC 3447 (RFC 3447, 2003) which describes how to use RSA in Internet protocols) with the SHA-1 hash algorithm used for calculating the digest. The message is verified by first verifying the CGA, then verifying the signature. The signature is checked by concatenating the 128-bit type tag with the message, minus the signature on the message. The test signature is then constructed exactly as described above and compared with the signature on the message. If both match, the message is authenticated.

It is important not to overestimate the applicability of CGAs to particular network security applications. CGAs prevent stealing and spoofing of IPv6 addresses, by binding the public key of the address owner cryptographically to the address through the hash. The address owner can then assert ownership and authorization to operate upon the address by signing signaling messages with the private key.

CGAs say nothing about whether the address owner can be trusted, since there is no requirement that the public key used to generate the address is certified. An attacker can generate a new CGA from a different subnet prefix and its public key or a victim's because the keys are not required to be certified. The attacker cannot, however, impersonate someone else's address because it would need to find a collision with *Hash1*. Of course, if the attacker does attempt to impersonate a victim by using the victim's public key and a different subnet prefix, the attacker cannot sign the message because it does not have the victim's private key. If a CGA is generated using a certified key, the certificate can provide trust verification for additional authorization on particular operations, as is the case for certified routers in SEND.

The CGA algorithm includes a few mechanisms for increasing resistance to collision searching attacks. By including the subnet prefix into the hash, an attacker is prevented from precomputing attack addresses with different subnet prefixes. An attacker must create a separate attack database for each subnet prefix. Link local addresses, which use the same subnet prefix regardless of the link, are still at risk from precomputation, however.

The *Sec* parameter provides another, more powerful mechanism. A *Sec* parameter greater than zero requires both the legitimate owner of the address and an attacker to search through the state space until the leftmost 2^{16Sec} bits are zero. This increases the number of operations required for an address generation and an attack by a factor of 2^{16Sec}. As a result, the cost of generating a CGA Parameters option binding the attacker's public key with the victim's address is increased from $O(2^{59})$ to $O(2^{59+Sec})$. Using a higher *Sec* value for link local address generation than is necessary for global unicast address generation can also help mitigate the higher threat to link local addresses. If N subnets must be protected against link local attack, then using a *Sec* value such that $2^{16Sec} > N$ reduces the threat to link local addresses to a point where it is no longer a concern. Since the primary attack on local subnet address generation and resolution is DoS, starting with a lower *Sec* value for these applications and moving higher only if collision attacks appear is a sensible strategy. This causes legitimate nodes to incur the extra computational effort only in a sufficiently realistic threat environment. If CGAs are used in other applications where stronger authentication or confidentiality are important goals, however, a higher *Sec* value may be necessary from the start.

The third protection for collision resistance is the inclusion of the collision count value into the calculation of *Hash1* and the limitation to three collisions. Normally, since the input values for a CGA are mostly random and providing the generating node's pseudorandom number generator is good, the probability of a collision is extremely low if the address space utilization is not densely packed. If a collision does occur, it is probably either a configuration error or a deliberate denial-of-service attack. If the number of collisions is not limited, an attacker that is doing a brute-force search to match a CGA can try different values for the collision count without repeating a brute-force search for the random modifier. The limitation on the number of collisions thereby increases the effectiveness of the *Sec* hash extension in preventing collision attacks.

Finally, RFC 3972 recommends that the RSA public key used to generate a CGA must be at least 384 bits long. This is too short for most RSA applications (1028 bits is the currently recommended shortest length), but at the time the CGA specification was finalized, this was considered sufficiently long enough that integer factoring attacks were impractical. Nowadays, a longer key is probably better, due to improvements in integer factoring attacks and the newest cryptanalysis results on collisions for SHA-1 described in Chapter 3. The purpose of appending the type tag messages before calculating the message digest is to prevent attackers from creating a CGA from another public key then replaying signed messages from another protocol. By including the type tag, signatures and keys are bound to a particular protocol. In addition, the RSA key used to generate a CGA must not be used for encryption, since RSA has a cryptographic vulnerability if the same key is used for both signing and encryption.

SEND protocol

The SEND protocol described in RFC 3971 is an extension of Neighbor Discovery and address autoconfiguration that utilizes CGAs to provide protection against DoS attacks associated with stealing or spoofing a node's IP address or spoofing a router. The SEND protocol consists of two main parts:[3]

- Use of CGAs for securing address resolution and address autoconfiguration. This involves the Neighbor Discovery NS and NA messages.
- Use of CGAs and router certificates for securing router discovery. This involves the Neighbor Discovery RS and RA messages, and adds a new message pair, Certification Path Solicitation (CPS) and Certificate Path Advertisement (CPA), for discovering the certification paths associated with routers' public key certificates.

SEND does not alter the basic Neighbor Discovery protocol for address resolution and address autoconfiguration. Instead, SEND requires three additional options in all NS and NA messages and all RS messages unless they are sent from the unspecified address (designated "::" in IPv6). These additional options are:

- the CGA Parameters option data structure described above in the section on CGAs, with some additional fields to make it a proper Neighbor Discovery option;
- the RSA Signature option;
- a Timestamp or Nonce option, used to prevent replay attacks.

---

[3] There is also an additional Neighbor Discovery message, called Redirect, which allows a node to redirect routing to its IPv6 address. This message can also be secured with SEND but for purposes of simplicity, we ignore it here. Please see RFC 3971 for a full description of the protocol.

The CGA itself appears in different places depending on the message. If the NS message is used for duplicate address detection, i.e. a node wants to configure the CGA on its interface and is trying to determine if any other node has the address configured, the CGA is the tentative address as described above in Figure 5.3. The CGA appears in a Target Address option of the multicast Neighbor Solicitation message. For other messages, sent after the address has been configured on the network interface, the CGA appears in the IPv6 header as the source address of the message. When a terminal verifies a message protected with SEND, the CGA is always verified prior to verifying the signature, since signature verification is more time consuming.

The CGA Parameters option data structure is constructed as described above after the generation of the CGA. To convert the data structure into valid Neighbor Discovery option format, 4 bytes are appended to the front of the option. These bytes are (in the order which they are appended):

- the IPv6 Neighbor Discovery option Type field for the CGA Parameters option (11);
- the Length of the option, in bytes, from the beginning of the option to the end;
- the Number of Pad Bytes field;
- a Reserved field set to zero by the sender and ignored by the receiver.

After the CGA Parameters option data structure, the option is padded out to an even multiple of 8 by adding zero bytes. The number of padding bytes is included in the pad length field at the beginning of the message.

Figure 5.6 illustrates the format of the RSA Signature option. As with the CGA Parameters option, it has a Type (12), Length, and Reserved field at the beginning. The number of pad bytes in this case is calculated from the total length and the lengths of the other fields rather than being explicitly stated. The Key Hash field includes the 16 most significant (leftmost) bytes of the public key hash used to construct the signature and the CGA. The receiver associates the signature to a particular key using the Key Hash field, in the event the receiver has the key cached. The Digital Signature field contains a variable-length RSA signature of the message constructed as described below and encoded in a particular format (PKCS#1 v1.5). After the digital signature, the option is padded out to an even multiple of 8 bytes using zeros.

To construct the digital signature, the sender first concatenates the following:

- the 128-bit CGA message type tag for SEND (the exact value is 0x 086F CA5E 10B2 00C9 9C8C E001 6427 7C08);
- the 128-bit IPv6 source address;
- the 128-bit IPv6 destination address;
- the 8-bit Type, 8-bit Code, and 16-bit Checksum fields from the Neighbor Discovery ICMP transport header;
- the Neighbor Discovery message header, from the byte after the ICMP transport header up to but not including the options;
- all the options up to but not including the RSA Signature option.

The concatenation of these values is digested by SHA-1 to form the message digest and the signature is calculated using the RSASSA-PKCS1 v1.5 algorithm as defined in

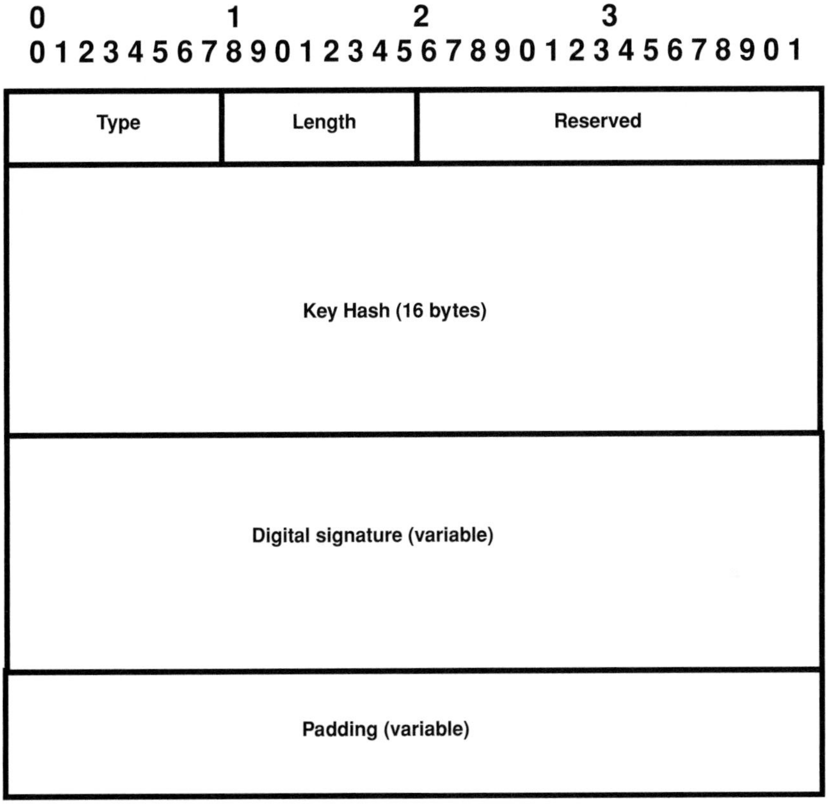

```
0                   1                   2                   3
0 1 2 3 4 5 6 7 8 9 0 1 2 3 4 5 6 7 8 9 0 1 2 3 4 5 6 7 8 9 0 1
```

Type	Length	Reserved

Key Hash (16 bytes)

Digital signature (variable)

Padding (variable)

Figure 5.6 RSA Signature option

the RSA Encryption Standard Version 2.1. The signature is then inserted into an RSA Signature option along with the other required fields and the RSA Signature option is appended at the end of the Neighbor Discovery message. The receiver calculates the test signature by removing the RSA option and using the remaining fields to calculate the signature, then comparing the test signature with the signature on the message. If the two match, then the signature is validated.

The Timestamp option prevents unsolicited NAs and RAs from being used in replay attacks. The receiving node checks the timestamp value and if the difference between the timestamp and the receive time is outside a particular network delta (default value is 5 minutes), the message is suspect. There are a few additional tests that can be applied to determine whether to accept or reject the message and exactly what operations to perform if the message is accepted; these are described in detail in RFC 3971. Similarly, the Nonce option prevents RSs and NSs from being used in replay attacks. When a solicitation is sent, the sending node includes a Nonce option with a randomly generated nonce. Nodes replying to the solicitation include the same Nonce option so the soliciting node can match the request to the reply. If there is no outstanding request, the soliciting node can identify the message as a replay attack and drop it. The Timestamp and Nonce options contain a Type field (13 for Timestamp, 14 for Nonce), a Length field, a Reserved field (Timestamp option only), then the Nonce or Timestamp. The Nonce option must be at least 6 bytes out of an at least 8 byte long random number. The Timestamp option is 8 bytes and the time is formatted according to the number of seconds since January 1, 1970 00:00 UTC in fixed-point format; more details are available in the specification.

For router discovery, SEND provides some additional protections. Unlike address resolution and duplicate address detection, security of last hop router discovery requires a mechanism whereby the node soliciting routing service can verify the authorization to route of a node advertising routing service. SEND requires the keys used to sign RAs to be certified; that is, the public keys must be accompanied by a public key certificate containing the key and signed by a certification authority. This requirement establishes the basis for a node soliciting routing service to trust the node offering routing service, through the trust intermediary of the certification authority.

While one level of authorization derives from the certificate conferred by a mutually trusted certification authority, SEND provides another level of authorization whereby a network operator can constrain the subnet prefixes routers are authorized to advertise. This finer-grained level of authorization can be used for various purposes by network operators. For example, two network operators that are sharing the same underlying physical infrastructure can configure two different access routers on the same last hop subnet with two sets of prefixes and certificates from their certification authorities authorizing access. A customer node will then only select the access router with a certificate matching the network operator with which the customer has an account. SEND defines a specific router authorization certificate profile for an X.509 router certificate, including an extension that lists the address prefixes delegated by the network operator to the router, which the router is authorized to route. The extension is contained in the *addressesOrRanges* attribute. Deployment of this extension is optional, since managing certified subnet prefixes might be beyond the technical means of some network operators. Router certificates authorizing basic routing capability, however, are required.

In order for a node to verify a router certificate, it must be able to verify the certification path from a certificate signed by a common trust anchor to the router certificate. For this purpose, the node maintains a certificate cache of well-known trust anchor certificates, much as is the case for TLS authentication of https sessions in Web browsers. This model of authentication and authorization – server provides authentication and authorization to client via a cache of trust anchor certificates and a certification path – has proven successful because it does not require bidirectional certification. The TLS authentication model avoids having to provision every IPv6 node with an individual certificate; instead, the cache of trust anchor certificates is held in common with other IPv6 nodes and can be preprovisioned with the operating system, as is done with Web browsers.

If a node receives an RA signed by a router for which it does not have a certificate, the node requests the router's certification path using Certification Path Solicitation (CPS) message defined in RFC 3971. The soliciting node includes the identifying names of the trust anchors which are in the node's certificate cache. The router replies with a sequence of Certification Path Advertisement messages, one message per certificate with the common trust anchor's certificate first and the router's certificate last. Upon reception of each CPA message, the requesting node verifies the certificate, until, when the final CPA message is received, the router's certificate is verified.

The CPS/CPA messages are sent unicast between the node and the router using the ICMP transport protocol. Because the certification path is required to bootstrap trust between the router and requesting node, it is not possible to certify the CPS/CPA messages, but the messages can still originate at a CGA constructed using the router's certified public key and be signed with the router's certified public key. This provides the requesting node with assurance that the messages did originate with the node whose routing credentials are being checked and that the messages were not modified in transit.

After the path has been verified, the requesting node also needs to check on the validity of the certificates using a certificate revocation list check. Since this operation typically requires routing access to the Internet, it cannot be completed before the router's certificate is validated. If any problem occurs during the certificate revocation list check – for example, messages are lost or the check is delayed – the requesting node should exercise caution. If the access router was compromised, it may be disrupting the check to avoid detection. The requesting node should, in that case, select a different access router if one is available and report the problem to the network operator. Nodes are allowed to cache certification paths, but periodic certificate revocation list checks are still required.

While SEND provides protection against the threats to address resolution, address autoconfiguration, and router discovery at the IPv6 layer, it does not compensate for an insecure link layer. In particular, on 802.11, it is possible for nodes to spoof their link address. This would allow a node to send out a NA on an unsecured link layer with the frame's link layer address set to the source address of a victim, a valid CGA address and a valid signature constructed by the attacker, and a Target Link Layer Address option corresponding to the victim. This would look to all nodes like a valid SEND-secured NA, and the attacker could then arrange for a DoS attack in which a high-volume traffic stream bombards the victim. This can be prevented on 802.11 by using 802.1x/802.11e and port-based access control, as described in Chapter 4. Port-based access control binds a port on the 802.11 access point to a particular link layer address with a particular shared key used to calculate encryption and authentication on the frame. If an authenticated attacker attempts to spoof the link layer address of a victim, the access point will retrieve the victim's key and not the attacker's, and so will be unable to validate and decrypt the frame. This will cause the frame to be dropped.

5.6   Security protocols for Local Subnet Configuration Server access

Security between the Local Subnet Configuration Server and a configuring Basic IP Node involves interfaces LCS1 and LCS2 in the architecture as shown in Figure 5.4. The protocol for Local Subnet Configuration Server access in IP networks is DHCP, DHCPv4 for IPv4 and DHCPv6 for IPv6. As described in Table 5.4, one protocol is the DHCP Authentication Option described in RFC 3118. This option is available in either IPv4 or IPv6, and is included with the DHCP message. Another protocol is IPsec authentication, either Authentication Header (AH), described in RFC 4302 (RFC 4302, 2005), or Encapsulating Security Payload (ESP), described in RFC 4303 (RFC 4303, 2005). IPsec is used at the IP layer between two different nodes. IPsec authentication is feasible only in IPv6, since IPv6 nodes are required to deploy IPsec in order to be specification-compliant. In IPv4, IPsec is optional and is typically used only for VPN access. For the LCS2 interface – credential provisioning and key exchange – the primary recommendation in RFC 3118 is to use manual key provisioning. Manual key provisioning can be used for IPsec, though dynamic key provisioning is also possible using IKEv1, described in RFC 2409 (RFC 2409, 1998), and IKEv2, described in RFC 4306 (RFC 4306, 2005).

While standardized mechanisms have been developed for DHCP authentication, in reality almost no one deploys them. Because the DHCP Authentication option depends on manual key provisioning, the administrative task of provisioning all wireless terminals in a network for the specific service of DHCP authentication is fairly daunting. While some network access authentication systems deploy shared keys, the AAA server and wireless terminal share a single long-term secret which is then used to derive service-specific and session-specific keys during the network access authentication transaction. Service-specific manual key provisioning is feasible only for small networks with very few users. In theory, shared key provisioning for DHCP authentication could be automated if the shared key is derived during link layer network access control, since link layer network access control typically does not involve IP traffic from the wireless terminal and thus does not require the terminal to be configured for the local subnet. In practice, there are no specifications about how to do this.

Despite the lack of deployment for authentication mechanisms protecting DHCP and lack of scalable credential and key management, we briefly describe the DHCP Authentication option below. IPsec and IKE are extensively discussed in Chapter 6, since they are the foundation of security for IPv6 mobility management.

5.6.1   DHCP Authentication option

Figure 5.7 illustrates the format of the DHCP Authentication option.

The Code field contains the DHCP option code for the Authentication Option (90). The Length field contains the length of the option starting with the Protocol field (i.e. ignoring the Code and Length fields) in bytes. The Protocol field specifies what protocol is used for calculating and verifying the authentication information, the Algorithm field allows further refinements in the authentication protocol. The RDM (Replay Detection Method) field defines the anti-replay detection algorithm. The Replay Detection Information field contains the replay detection information, calculated using the specified Replay Detection Method. The Authentication Information field contains the authenticator calculated according to the Protocol and Algorithm fields.

RFC 3118 defines a single replay detection method having code 0. The Replay Detection Information field must be set to a monotonically increasing counter. The specification recommends using a timestamp calculated in Network Time Protocol (NTP) format. Other replay detection algorithms can be defined and new code points registered with the Internet Assigned Numbers Authority (IANA).

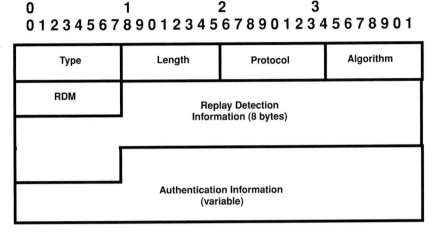

Figure 5.7 DHCP Authentication option format

Two authentication algorithms are defined in RFC 3118, Configuration Token (Protocol code 0, Algorithm code 0) and Delayed Authentication (Protocol code 1). The Configuration Token protocol specifies that the sender and receiver share an opaque token which the sender inserts into DHCP messages. The receiver matches the token against its database of tokens for authorized users and authenticates the message if a token from the database matches. The intent of the protocol is that the token works something like a password. Other types of token-based protocols such as one-time passwords should be defined as separate protocols. Needless to say, cryptographic assurances are lacking in this protocol.

In the Delayed Authentication protocol, the sender and receiver share a key and use the shared key to compute a message authentication code for the message. RFC 3118 defines a single algorithm for calculating the message authentication code, namely HMAC-MD5, with Algorithm code 1. The message authentication code is calculated over the entire DHCP message, including the header and options, setting the MAC field in the Authentication Information field to zeros for computation. The Delayed Authentication protocol defines two formatted Authentication Information suboptions, one for DHCPDISCOVER and DHCPINFORM, and one for DHCPOFFER, DHCPREQUEST and DHCPACK. In the first, there is no authentication information, while in the second, the Authentication Information field contains a 32-bit key name (called a secret ID in RFC 3118) and 128-bit HMAC-MD5 message authentication code.

In the Delayed Authentication protocol, the client indicates it wants to use Delayed Authentication by setting the Protocol and Algorithm fields to 1 in the option sent with the DHCPDISCOVER message and also includes a DHCP Client Identifier option to uniquely identify itself to the server. The server returns DHCPOFFER messages with the requested Authentication Information field in the option set to the key name and HMAC-SHA1 message authentication code. The client validates the message authentication codes of messages returned from DHCP servers, using the key specified by the key name in each message, and selects one that validates. The client then replies with a DHCPREQUEST message that is protected in the same manner, the server replies with DHCPACK also protected with a message authentication code.

5.7   Summary

Local IP subnet configuration and address resolution are critical operations for wireless terminals. When a wireless terminal initially enters the network or after a handover to a new access point in a new subnet, local IP subnet configuration is necessary in order for the terminal to obtain a new access router and new IP address so that it can continue receiving IP routing service. Address resolution is necessary so that the access router or terminals on the local subnet can resolve the terminal's IP address to a link address for delivery of packets on the last hop. While these operations are important for fixed terminals too, wireless terminals are mobile and tend to use these operations more frequently because wireless terminals change IP subnets more frequently. Most fixed terminals never change to a new IP subnet after they initially connect. Security for local IP subnet configuration addresses different threats and does not replace network access control and security at the link layer.

A security architecture for local IP subnet configuration and address resolution requires accommodating the existing protocols for performing local IP subnet configuration and address resolution. The IP protocols for IP subnet configuration and address resolution are different in IPv4 and IPv6. In IPv4, a link layer protocol, Address Resolution Protocol (ARP) is used for address resolution. Local IP subnet configuration is accomplished using a server-based protocol, Dynamic Host Configuration Protocol, DHCP. For IPv6, both address resolution and subnet configuration are accomplished using Neighbor Discovery, an IP layer protocol. IPv6 also supports a version of DHCP that allows server-based address configuration for globally routable IPv6 unicast addresses.

The primary threat to IP subnet configuration and address resolution is address spoofing. In address spoofing, an attacker claims the address of a victim, through spoofing address resolution messages. This allows the attacker to intercept the victim's traffic and do with that traffic as they see fit. Another similar attack is spoofing a router during access router discovery. This allows the attacker access to traffic from all nodes on the subnet. Attacks on address configuration allow an attacker to mount a DoS attack on a node. If the node is unable to obtain an IP address due to the attacker's interference, the node becomes unable to obtain IP routing service to the Internet.

The architecture we defined for security of local IP subnet configuration and address resolution involves four different interfaces. The BN1 interface is between two different IP nodes, and involves authentication for address resolution to prevent basic address spoofing. The AR1 interface is between the access router and any other IP node on the subnet, and it involves authentication of router advertisements to prevent disruption of router discovery. The LCS1 and LCS2 interfaces are for server-based subnet configuration. These interfaces are between any IP node and the local subnet configuration server.

The LCS1 interface supports basic authentication of client-server messages, the LCS2 interface supports credential provisioning and key distribution.

The actual security for the configuration and address resolution protocols in IP networks vary in quality. Since most IPv4 networks were deployed before security was considered an important design criterion and since configuration and address resolution are fundamental operations that were deployed quite early, the IPv4 protocols for these operations tend to have inadequate cryptographic protection. ARP, for example, has no cryptographic protection and depends on certain complicated deployment measures to thwart attacks. DHCP has the DHCP Authentication Option, but because credential provisioning and key distribution mechanisms were neglected, the DHCP Authentication Option is not widely deployed. In contrast, security for subnet configuration and address resolution was added to IPv6 Neighbor Discovery before widespread deployment. Authentication for address resolution and address autoconfiguration utilizes Cryptographically Generated Addresses (CGAs), a powerful technique based on public keys that allows an address and messages operating upon it to be cryptographically tied to the owner of a public key. Router discovery security in IPv6 is supported by public key certificates for routers, which allow a node to trace back the router to a trusted certification authority certifying the router's authority to route. Server-based address security in IPv6 using DHCP can utilize the same Authentication option as in IPv4, or additionally IPsec for authentication and IKE for key provisioning. IPsec/IKE is an option for IPv6 security because these protocols are required to be deployed on IPv6 nodes, for IPv4 they are optional.

6   Security for global IP mobility

Once a wireless terminal has cleared network access control, obtained an IP address on the local subnet, and has routing service for IP packets between the terminal and the network, the terminal has access to the higher-level services available on the global Internet – Web pages, IP telephony, streaming video and the like. From the point of view of routing and packet delivery service, a wireless terminal is no different than a wired terminal. A desktop PC connected to the Internet through DSL must go through a similar process to get Internet access as a wireless terminal and the resulting routing and packet delivery service is basically the same. Unlike the user of a desktop PC, however, the user of a wireless terminal is free to move the terminal to a new location. Such a movement may cross an invisible line in the access network topology between a geographical area where the current IP address continues to provide packet delivery service and where the address stops functioning. In other words, the terminal moves from one IP subnet to another causing IP handover to occur.

If the user's mobility patterns conform to the nomadic usage model discussed in Chapter 4, then starting network access control and local IP subnet configuration from the beginning are adequate for initiating routing and packet delivery service in the new subnet. The user has no expectation that sessions started in the old subnet continue in the new subnet, because all sessions are closed when the wireless terminal disconnects from the old subnet. No existing session traffic is disrupted by the move. If the mobility pattern conforms to the full mobility model, however, the user has a reasonable expectation that higher-level services continue to operate while the terminal is moving. The user's expectation generates a requirement for session continuity, that the network continue delivering traffic in the new subnet for sessions that were started in the old subnet. The various mechanisms used to solve this problem at the architectural, system, and protocol levels are called *IP mobility*.

In the next section, we briefly review existing standards in IP mobility support, specifically Mobile IP. Two versions of Mobile IP have been standardized: Mobile IPv4 (RFC 3344, 2002), which supports IPv4, and Mobile IPv6 (RFC 3775, 2004), which supports IPv6. In this chapter, we focus on Mobile IPv6 because the security architecture in Mobile IPv6 cleanly separates the home network functions, access network functions, and wireless terminal functions, and the protocol itself is therefore more modular. We then develop a functional architecture for Mobile IPv6 security, designating functions and interfaces on which protocols are required. As in previous chapters, the functional architecture closely follows the Mobile IPv6 architecture that has been standardized.

Finally, we briefly review the security protocols on the interfaces in the Mobile IPv6 architecture, including IP Security (IPsec), the Internet Key Exchange (IKE) version 2 protocol, and the return routability protocol.

6.1   Review of IP mobility architecture and protocols

As discussed in Chapter 5, wireless IP networks are deployed as a collection of wireless access points that route to and from a wired access network and the global Internet through access routers. Each access router controls a different wireless stub subnet, to which wireless terminals connect. A wireless terminal incurs a problem when it moves between one stub subnet and another. The problem is that an IP address which works fine in one subnet does not work when the wireless terminal moves to a new subnet. The reason is because the IP address functions as a *routing locator*, directing the routing system in the network where to route a packet with that IP address as the destination address. If a wireless terminal moves to a new subnet, packets sent by the correspondent node with which the wireless terminal has a session are forwarded to the old access router and are dropped, because the terminal is no longer there. In order for the packets to reach the terminal, the destination address must be the wireless terminal's address in the new subnet.

A seemingly straightforward solution is to just have the wireless terminal send its new address to the correspondent. Packets in transit while the address is changing would be lost, but after an initial transient loss, the wireless terminal would receive packets at the new address. The problem with this scheme is that, besides functioning as a routing locator, the IP address together with the port number also functions as an *end node identifier* for certain transport protocols like TCP. The end node identifier identifies a session with a particular end host, and switching the address would cause the session to drop on the correspondent node. This basic architectural problem with IP addresses and mobility is called the *identity/locator problem*, and it is the reason why IP mobility management protocols are needed. Basic IP routing does not support mobility.

6.1.1   Mobile IP architectural overview

There have been a variety of protocols proposed to solve the identity/locator problem over the years, both on a research basis and as standards. However, the protocol family that has had the most development as a standard is Mobile IP. Mobile IP provides mobility support by anchoring the routing for wireless terminals at a mobility management router in the home network, called the *home agent*. Figure 6.1 contains an overview of Mobile IP routing through the home agent. Initially, all packets exchanged with correspondent nodes are routed through the home agent, which does not move when the wireless terminal moves.

At a high level, the steps involved in IP mobility management with Mobile IP are the following:

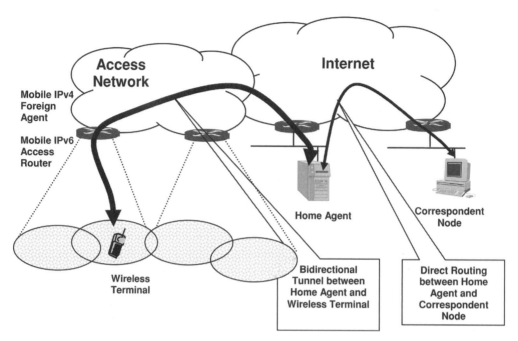

Figure 6.1 Mobile IP routing

- The wireless terminal obtains an address in a subnet served by its home agent, its *home address*, and uses that address as a source address when exchanging traffic with a correspondent node. The source address acts as the end node identifier.
- The correspondent sends traffic to the home address rather than directly to the wireless terminal.
- The traffic is intercepted by the home agent, and tunneled to the wireless terminal. The tunneling is accomplished by encapsulating packets from the correspondent node in packets with a destination address in the wireless terminal's local subnet, called the *care-of address*, and forwarding them to the wireless terminal. The care-of address acts as the routing locator.
- The wireless terminal also tunnels packets back to the correspondent node through the home agent, using the care-of address as the source address. This ensures that the wireless terminal's packets have a topologically correct source address in the access network. The home agent decapsulates the packets and forwards them to the correspondent node.
- When the wireless terminal moves to a new subnet, it updates the binding between the home address and the care-of address at the home agent, so the home agent knows where to tunnel packets.

The home agent and wireless terminal manage a bidirectional tunnel between them, ensuring that packets to and from the wireless terminal are properly routed.

This basic architecture is used by Mobile IP for both versions of the IP protocol: Mobile IPv4 and Mobile IPv6. However, there are a few differences. Mobile IPv4 also includes a specialized last hop router in the access network, called a *foreign agent* where the care-of address for the wireless terminal is actually located. If a foreign agent is present, the home agent manages a bidirectional tunnel with the foreign agent rather than the wireless terminal, though it is possible for the wireless terminal to manage the tunnel itself if there is no foreign agent. If the foreign agent manages the tunnel, the care-of address is located on the foreign agent rather than the wireless terminal. If there is no foreign agent, the care-of address is co-located on the wireless terminal. The foreign agent has a large impact on the security architecture for Mobile IPv4, as we will see later in the chapter.

Mobile IPv6 also includes some protocol support for route optimization, which is not included in Mobile IPv4. Since packets for Mobile IP hosts are always routed indirectly through the home agent, the latency in packet delivery could be potentially much worse than for direct delivery. In the worst case, the wireless terminal and its correspondent are in exactly the same wireless subnet and the home agent is located on another continent. Packets between the two then need to cross an ocean and return, whereas, if the wireless terminal were not using Mobile IP, the packets would simply go to the nearby access router and back across the wireless link. Because the protocol for route optimization requires fundamental changes in the IP stack, route optimization was not introduced in Mobile IPv4. IPv4 is already so widely deployed that the likelihood of deployment for a fundamental change in the stack was deemed too small to justify the work.

6.1.2   Mobile IP interfaces and protocols

There are four interfaces in the Mobile IP architecture, shown in Figure 6.2:

- For both Mobile IPv4 and Mobile IPv6, a *binding management interface* between the node holding the care-of address and the home agent supports a protocol that allows the node holding the care-of address to manage the binding between the home address and the care-of address. In Mobile IPv4, the node holding the care-of address is either the foreign agent or the wireless terminal, while in Mobile IPv6, the node holding the care-of address is always the wireless terminal. In Mobile IPv4, the binding management is called *registration*, while in Mobile IPv6 it is called *binding update*.
- For both Mobile IPv4 and Mobile IPv6, a *movement detection* interface between the wireless terminal and the foreign agent (for Mobile IPv4) or the access router (for Mobile IPv6) supports a protocol that allows the wireless terminal to detect when it has moved to a new subnet managed by a new last hop router. In Mobile IPv4, the interface also supports a registration protocol allowing the terminal to obtain a new care-of address. In Mobile IPv6, this function is handled by standard IPv6 subnet configuration protocols, described in Chapter 5, and is not part of the Mobile IPv6 specification.
- In Mobile IPv6, the *route optimization* interface between the wireless terminal and correspondent node supports a protocol that allows the wireless terminal to optimize routing, allowing packets to be routed directly between the correspondent node and

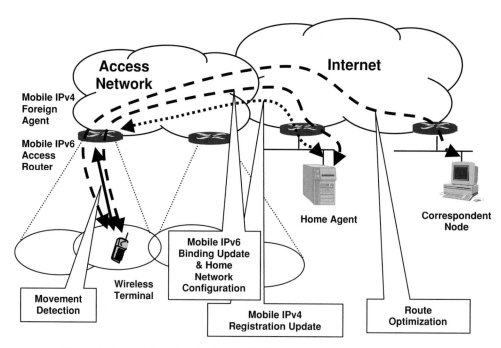

Figure 6.2 Network interfaces in the Mobile IP architecture

wireless terminal rather than through the home agent. This interface is absent in Mobile IPv4 because route optimization was not developed for it.

- Mobile IPv6 has a *remote home subnet configuration* interface between the wireless terminal and the home agent. The remote home subnet configuration interface supports a protocol that allows the wireless terminal to discover a home agent and to obtain information on the home subnet prefixes when it is connected to a remote subnet not part of the subnet where the home agent is located. These operations are typically done on the local IP subnet using the IPv6 Neighbor Discovery protocol discussed in Chapter 5. Mobile IPv6 provides similar functionality for remote hosts.

The binding management protocol between the wireless terminal and the home agent in both Mobile IPv4 and Mobile IPv6 and the route optimization protocol between the wireless terminal and correspondent node in Mobile IPv6 are both request/response protocols. The wireless terminal sends a request to the other end to establish a new binding between the two IP addresses, to change an existing binding to a new care-of address, or to delete a binding. The other side responds with a reply indicating if the response was successful. The movement detection protocol in Mobile IPv4 and the remote home subnet configuration protocol in Mobile IPv6 support a request/response style, but they also support an unsolicited style. The unsolicited style allows the network-side entity – the foreign agent for the movement detection protocol or the home agent for the remote home subnet configuration protocol – to send an unsolicited message to the wireless terminal. The home agent discovery message on the remote home subnet configuration interface is always a request/response message, however.

6.2   Threats to Mobile IP security

Threats to Mobile IP occur on each of the four interfaces. The threats to the binding management interface between the home agent and the wireless terminal and to the interface between the wireless terminal and the correspondent node are similar since both interfaces support request/response protocols. An important difference is that the wireless terminal is assumed to have a trust relationship with the home agent while no such trust relationship is assumed between the wireless terminal and a correspondent node. The home agent must verify the wireless terminal's identity before establishing the initial binding. This trust relationship may be a business relationship, as when the owner of the wireless terminal is a customer of a home network service provider, or it may be an authorized user relationship, as when the owner of the wireless terminal is an employee in an enterprise that has deployed Mobile IP service for wirelessly connecting to an enterprise network. On the other hand, since the correspondent node could be any random node in the Internet, no explicit trust relationship can be assumed between the wireless terminal and the correspondent node because the Internet as a whole does not support identity verification between two random nodes.

This rest of this section discusses threats on the home agent to wireless terminal interface for binding management and remote configuration, and threats on the wireless terminal to correspondent node interface for route optimization. Threats on the movement detection interface between the wireless terminal and the access router were covered in Chapter 5 and, in any case, this interface is not part of Mobile IPv6, which is the main focus of discussion in this chapter.

6.2.1   Threats to the binding management and remote home subnet configuration interfaces

There are four basic attacks on the binding management interface between the wireless terminal and the home agent:

- The chief threat to binding management is that an attacker could send an unauthorized binding update message to the home agent and cause traffic for the victim to be redirected to a target of the attacker's choice. This kind of attack could be used to deny the victim routing service if the binding is deleted, to snoop traffic if the new care-of address is the attacker's, or to bombard another unsuspecting victim with a high-volume packet stream that was not expected (a so-called "bombing attack") if the new care-of address is that of a third party, unsuspecting victim.
- An even more insidious attack is when an authenticated wireless terminal sends a binding update to the home agent changing the binding on a home address for which it is not authorized.
- An attack is also possible on the wireless terminal from the home agent side. A reply to a binding update message indicating that the binding was not successful could lead the terminal to conclude that IP mobility service was not available, causing a denial of service.

- As with any request/response protocol, replay attacks are possible even if the binding update signaling is protected with data origin authentication. The attacker records an authentic binding update message, and then replays it later to the home agent when the wireless terminal has moved to a new care-of address. This causes the home agent to move the routing back to an old care-of address, effectively denying packet service to the wireless terminal.

The security services that counter the threats above are:

- Bidirectional identity authentication between the wireless terminal and home agent when the wireless terminal initially tries to establish a binding counters the threat of an unauthorized terminal or rogue home agent. Identity authentication ensures that the terminal is authorized for IP mobility management and that the home agent is the authorized mobility agent for the wireless terminal.
- The initial identity authentication exchange also provides key management to establish a security association between the home agent and the wireless terminal. The key management provisions a key for the data origin authentication on the home agent and on the wireless terminal.
- Every instance of binding update signaling requires data origin authentication to ensure that the signaling originated with an authenticated wireless terminal authorized to change the routing for the home address. Similarly, every binding update reply requires data origin authentication to ensure that signaling originated with the wireless terminal's authorized home agent.
- In addition to data origin authentication, the home agent must also verify that the binding update originated from a wireless terminal that is authorized to change the requested home address. This eliminates the insider attack where an authorized terminal tries to change a binding for a home address that it is not authorized to change.

The wireless terminal may also be concerned about an attacker intercepting clear text tunneled data traffic to and from the home agent. Establishing confidentiality protection on tunneled data traffic ensures that an attacker cannot snoop the tunneled data. The home agent in this case functions like a virtual private network server. Confidentiality service is not strictly required, since the wireless terminal could also establish an end-to-end security service with the correspondent node if the nature of the traffic is sensitive. However, as we will see later in the chapter, the Mobile IPv6 return routability protocol signaling, which secures route optimization between the correspondent node and wireless terminal, requires confidentiality protection on the home agent to wireless terminal tunnel. This requirement evolves from the implementation of the return routability protocol rather than resulting from the architecture. Consequently, we do not include confidentiality into the architecture except as an optional service for wireless terminals that are concerned about interception.

The remote home subnet configuration interface is subject to the following attacks:

- The home agent discovery message could be intercepted by a bogus home agent. A bogus reply causes the wireless terminal to set up a binding with an attacker posing as its home agent. The wireless terminal's traffic is then subject to examination and redirection by the attacker.

- The wireless terminal could send a home subnet prefix discovery request to an attacker posing as a home agent. Alternatively, the attacker could send an unsolicited home subnet prefix advertisement posing as a home agent. The wireless terminal then uses that prefix to autoconfigure a home address but the address is rejected by the actual home agent, causing denial of service.
- An attacker uses the home subnet prefix discovery mechanism to discover interesting information about the internal topology of the home network. This information then helps the attacker to optimize attack patterns on routers and servers in the home network.

The security services used to mitigate the attacks on home subnet prefix discovery are the following:

- As with binding management, the home subnet prefix discovery request and reply messages must be protected with data origin authentication, so that both the wireless terminal and the home agent can verify the identity of the sender.
- The mobile prefix traffic may need to be confidentiality protected if the home network operator is concerned about an attacker using intercepted home network prefix discovery messages to discover information about the home network topology. If confidentiality protection is desired, a security association for confidentiality protection, including the keys for performing encryption, must be established at the time the initial terminal and home agent identities are mutually authenticated.

Determining what, if any, security services are required to mitigate attacks on home agent discovery is a little trickier. Clearly some procedure is required to ensure that the wireless terminal ends up talking to an authenticated home agent. However, since home agent authentication is a necessary step prior to binding initialization and mobile prefix discovery anyway, the home agent discovery message need not be authenticated. At worst, an attacker spoofing home agent discovery could spoof the wireless terminal into attempting to set up a security association but the attempt would fail once the wireless terminal determined that the presumptive home agent was unauthorized. Like any other IP protocol, this kind of attack could be used repeatedly to try to deny service to the wireless terminal, but at some point, the wireless terminal can simply give up and ignore the home agent discovery replies, concluding that it is under attack and that it is not going to get a useful answer. The terminal still may be able to do useful work with its care-of address, and might try to establish mobility management service later.

6.2.2   Threats to the route optimization interface

Because the correspondent node part of the route optimization protocol runs on every IPv6 node, attacks on the route optimization interface can be perpetrated against any IPv6 node, and the solutions need to be incorporated into IPv6 stacks even for hosts which are not mobile. As with the wireless terminal to home agent interfaces, the primary threats to the route optimization are unauthorized traffic redirection and replay attacks. The required security services are the same as those on the wireless terminal to home agent interfaces.

DoS attacks are an additional problem on this interface. An attacker can induce any IPv6 node to initiate binding updates with thousands of real or imaginary nodes, thereby causing the victim to waste resources on processing binding updates. This attack can be perpetrated even if the binding updates are authenticated. The victim IPv6 node can protect itself by limiting the number of binding updates it initiates, selectively dropping messages when the limit is reached, and, at some point, simply not doing route optimization, but these measures have consequences. The consequences deny route optimization to any correspondent node. These consequences can be mitigated by a variety of means: accepting Binding Update messages only from correspondent nodes with which the correspondent node under attack has some prior relationship, reserving more space in the host's binding cache, or aggressively returning to route optimization. DoS attacks are most effective against traffic that would suffer from the additional latency incurred by non-optimal routing, such as real-time audio and video, and for servers where the address is long-lived and published in the DNS.

Another attack that can be perpetrated on any IPv6 node with route optimization is the so-called "time shifting" attack. The attacker sends a spoofed binding update message to an IPv6 node even before the wireless terminal has established a binding, binding the attacker's own address as the new care-of address with the wireless terminal's home address. If the new binding is active when a correspondent wants to contact the wireless terminal, the correspondent's traffic routes to the wireless terminal through the attacker rather than through the home agent, allowing the attacker to become a man in the middle. Such attacks can be limited by limiting the lifetime of route optimization bindings.

6.3   Functional architecture for Mobile IP security

The functional entities for Mobile IP are dictated by the Mobile IP mobility management protocol. We capitalize the names of the functional entities to distinguish them from the network entities participating in the Mobile IP protocol. There are three functional entities:

- The Mobile Node, an IP host with a wireless interface capable of moving from one IP subnet to another.
- The Home Agent, a Mobile IP router in the Mobile Node's home network, which maintains a binding between the Mobile Node's home address in the home network and care-of address in the remote network and tunnels traffic to and from the Mobile Node.
- The Correspondent Node, which could effectively be any node on the Internet with which the Mobile Node is exchanging IP traffic.

The remainder of this section discusses the functional architecture, interfaces between the functional entities, the details of the functions, and a mapping of the functional

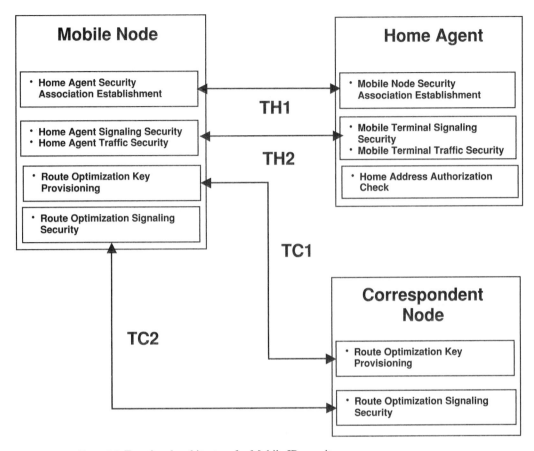

Figure 6.3 Functional architecture for Mobile IP security

architecture to the standardized protocols. In the development of the architecture for IP mobility management security, we initially do not distinguish between Mobile IPv4 and Mobile IPv6, reserving that discussion for later when we map the architecture to protocols

6.3.1   Functional architecture and interfaces

Figure 6.3 shows the security functional architecture for Mobile IP. Only the open network interfaces are shown. Internal programmatic interfaces may exist between the security functions and other functions, such as tunnel management between the home agent and wireless terminal.

There are 4 network interfaces in the Mobile IP security architecture:

- TH1 – the Mobile Node to Home Agent interface providing identity management and key management. This interface is responsible for setting up the initial security association between the Mobile Node and Home Agent, including key provisioning for security services on signaling and data traffic.

- TH2 – the Mobile Node to Home Agent interface providing security services for signaling and data traffic. These services include data origin authentication and replay protection for binding management and remote home subnet configuration signaling, and, optionally, data origin authentication and confidentiality protection for tunneled data traffic.
- TC1 – the Mobile Node to Correspondent Node interface providing key management for route optimization between the Mobile Node and Correspondent Node. TC1 supports the return routability protocol.
- TC2 – the Mobile Node to Correspondent Node interface providing data origin authentication and replay protection on route optimization signaling.

Note that there is no security interface between the Home Agent and Correspondent Node, because the Home Agent is really just acting as a router for the Mobile Node's traffic. The Home Agent is transparent as far as the Correspondent Node is concerned.

An alternative architecture would be to combine the TH1 and TH2 interfaces and the TC1 and TC2 interfaces so that there is only one interface each between the wireless terminal and home agent, and between the wireless terminal and correspondent node. This approach ties together identity management, key management and the security services on the actual signaling and data traffic. The resulting design is less modular. By keeping the interfaces and their implementing protocols separate, it is possible to replace one or the other protocol if new technologies become available, or if a security bug is found. In contrast, identity management and key management are intimately linked, since key provisioning requires authentication, so including both these functions on TH1 makes sense.

The descriptions of the functions in the next subsections assume that shared key cryptography is used between the Mobile Node and the Home Agent, and between the Mobile Node and the Correspondent Node for signaling and/or traffic security. This assumption is based on the actual design of both Mobile IPv4 and Mobile IPv6, which use shared key cryptography after key provisioning is complete.

6.3.2   Mobile Node functions

Table 6.1 contains a list of Mobile Node functions together with the security services they provide, the parameters for the functions, and the objects returned by the functions. The functions can be tied back to the threats through the security services and the discussion above. The following subsections describe the functions in more detail.

Home Agent Security Association Establishment function

The Home Agent Security Association Establishment function conducts the transaction that sets up a security association with the Home Agent. The parameters are the wireless terminal identity and credentials that are shared with the Home Agent longer term for identifying the Mobile Node. The function returns an indication of whether the security association was successfully established and the shared session keys for cryptographic operations on signaling and data traffic.

Table 6.1 Functions, parameters, and results for the Mobile terminal

Function	Security services	Parameters	Return
Home Agent Security Association Establishment	– Identity verification and key provisioning to establish a security association with the Home Agent – Security on signaling to establish a security association	– Mobile terminal identity – Long-term credentials shared with home network (e.g. username/ password, shared key, certified public key, etc.)	– Yes/no indication that the security association formation succeeded – Session keys shared with the Home Agent for data origin authentication and optionally encryption on signaling – Optionally session keys for data origin authentication and encryption on data traffic
Home Agent Signaling Security	– Data origin authentication and confidentiality protection on binding management and home subnet configuration signaling with the Home Agent	– Session keys for signaling security shared with the Home Agent – On sending, clear text message to be authenticated and optionally encrypted – On receiving, secured message to be verified and optionally decrypted	– On sending, the authenticated and optionally encrypted message – On receiving, the verified clear text message
Home Agent Traffic Security	– Data origin authentication and confidentiality protection on tunneled data traffic with the Home Agent	– Session keys for traffic security shared with the Home Agent – On sending, clear text packet to be authenticated and encrypted – On receiving, secured packet to be verfied and decrypted	– On sending, the authenticated and encrypted packet – On receiving, the verified clear text packet
Route Optimization Key Provisioning	– Key provisioning for route optimization security with the Correspondent Node	– Material for key generation	– Session key shared with the correspondent node for route optimization authentication
Route Optimization Signaling Security	– Data origin authentication on route optimization binding management with the Correspondent Node	– Session key shared with the correspondent node for route optimization authentication – On sending, message to be authenticated – On receiving, message to be verified	– On sending, authenticated message – On receiving, verified message

Home Agent Signaling Security function

The Home Agent Signaling Security function performs security on signaling between the Mobile Node and the Home Agent. The parameters include the session keys for cryptographic operations on signaling traffic. On sending, the clear text signaling message to be authenticated and, optionally, encrypted is a parameter. On receiving, the secured message to be verified and decrypted is a parameter. The function returns either the secured message on sending or the clear text, verified message on receiving.

Home Agent Traffic Security function

The Home Agent Traffic Security function performs security on data traffic routed on the tunnel between the Mobile Node and Home Agent. The parameters include the session keys for cryptographic operations on the data traffic. On sending, the clear text data packet to be authenticated and encrypted is a parameter. On receiving, the secured packet to be verified and decrypted is a parameter. The function returns either the secured packet on sending or the clear text message on receiving.

Route Optimization Key Provisioning function

The Route Optimization Key Provisioning function performs key provisioning for route optimization authentication with the Correspondent Node. Material for key provisioning is the parameter; the return is a session key for authenticating route optimization signaling traffic with the Correspondent Node.

Route Optimization Signaling Security function

The Route Optimization Signaling Security function performs authentication and verification on route optimization signaling traffic with the Correspondent Node. The parameters include the session key for route optimization authentication. On sending, the clear text route optimization signaling message for authentication is a parameter. On receiving, the secured message received from the Correspondent Node is a parameter. The function returns either the secured message on sending or the clear text message on receiving.

6.3.3   Home Agent functions

Table 6.2 contains a list of Home Agent functions together with the security services they provide, the parameters for the functions, and the objects returned by the functions. The following subsections describe the functions in more detail.

Mobile Node Security Association Establishment function

The Mobile Node Security Association Establishment function is the Home Agent counterpart to the Home Agent Security Association Establishment function on the Mobile Node. It conducts the transaction that sets up a security association with the Mobile Node. The parameters are the Home Agent identity and credentials that are shared with the Mobile Node longer term, such as a user name/password, used for

Table 6.2 Functions, parameters, and results for the Home Agent

Function	Security services	Parameters	Return
Mobile Node Security Association Establishment	– Identity verification and key provisioning to establish a security association with the Mobile Node – Security on signaling to establish a security association	– Home agent identity – Long-term credentials shared with Mobile Node (e.g. username/password, shared key, certified public key, etc.)	– Yes/no indication that the security association formation succeeded – Session keys shared with the Mobile Node for data origin authentication and optionally encryption on signaling – Optionally session keys for data origin authentication and encryption on data traffic
Mobile Node Signaling Security	– Data origin authentication and confidentiality protection on binding management and home subnet configuration signaling with the Mobile Node	– Session keys for signaling security shared with the Mobile Node – On sending, clear text message to be authenticated and optionally encrypted – On receiving, secured message to be verified and optionally decrypted	– On sending, the authenticated and optionally encrypted message – On receiving, the verified clear text message
Mobile Node Traffic Security	– Data origin authentication and confidentiality protection on tunneled data traffic with the Mobile Node	– Session keys for traffic security shared with the Mobile Node – On sending, clear text packet to be authenticated and encrypted – On receiving, secured packet to be verified and decrypted	– On sending, the authenticated and encrypted packet – On receiving, the verified clear text packet
Home Address Authorization Check	– Verify Mobile Node authorization to change home address	– Home address for which the binding is to change – Identity of the Mobile Node	– Yes/no indication of whether the Mobile Node is authorized to change the home address

identifying the Mobile Node. The function returns an indication of whether the security association was successfully established and the shared session keys for cryptographic operations on signaling and data traffic.

Mobile Node Signaling Security function

The Mobile Node Signaling Security function is the Home Agent counterpart of the Home Agent Signaling Security function on the Mobile Node. It performs security on signaling between the Mobile Node and Home Agent. The parameters include the session

Table 6.3 Functions, parameters, and results for the Correspondent Node

Function	Security services	Parameters	Return
Route Optimization Key Provisioning	– Key provisioning for route optimization security with the Mobile Node	– Material for key generation	– Session key shared with the mobile terminal for route optimization authentication
Route Optimization Signaling Security	– Data origin authentication on route optimization binding management with the Correspondent Node	– Session key shared with the mobile terminal for route optimization authentication – On sending, message to be authenticated – On receiving, message to be verified	– On sending, authenticated message – On receiving, verified message

keys for cryptographic operations on signaling traffic. On sending, the clear text signaling message to be authenticated and optionally encrypted is a parameter. On receiving, the secured message to be verified and decrypted is a parameter. The function returns either the secured message on sending or the clear text, verified message on receiving.

Mobile Node Traffic Security function

The Mobile Node Traffic Security function is the Home Agent counterpart to the Home Agent Traffic Security function on the Mobile Node. It performs security on data traffic routed on the tunnel between the Mobile Node and Home Agent. The parameters include the session keys for cryptographic operations on the data traffic. On sending, the clear text data packet to be authenticated and encrypted is a parameter. On receiving, the secured packet to be verified and decrypted is a parameter. The function returns either the secured packet on sending or the clear text packet on receiving.

Home Address Authorization Check function

The Home Address Authorization Check function is called by the binding management module to verify that the Mobile Node is authorized to change the requested home address. The function parameters are the home address for which the binding is to change and the identity of the Mobile Node. The function returns a yes/no indication whether the Mobile Node is authorized to change the address. Note that this function does not handle verification of the authorization on the binding update message, that is handled by the Mobile Node Signaling Security function.

6.3.4   Correspondent Node functions

Table 6.3 contains a list of Correspondent Node functions together with the security services they provide, the parameters for the functions, and the objects returned by the functions. The following subsections describe the functions in more detail.

Table 6.4 Mapping of interfaces to protocols for Mobile IPv4 with Foreign Agent

	Functional elements			Protocols on interfaces			
	Supplicant	Authenticator	Account Authority	N1	N2	N3	N4
Mobile IPv4 with Foreign Agent	Terminal	Foreign Agent	Home Agent AAA Server	RFC 4721 RFC 2794 RFC 3957	RFC 4721 RFC 2794 RFC 3957	RFC 4721 RFC 2794 RFC 3957 Possibly Radius	<None>

Route Optimization Key Provisioning function

The Route Optimization Key Provisioning function is the counterpart to the Route Optimization Key Provisioning function on the Mobile Node. The parameters and return value are the same.

Route Optimization Signaling Security function

The Route Optimization Signaling Security function is the counterpart to the Route Optimization Signaling Security function on the Mobile Node. The parameters and return value are the same.

6.3.5 Taxonomy of deployed systems

The presence of the foreign agent in the Mobile IPv4 mobility management architecture led to a security architecture that is substantially different from the architecture described above. The original security architecture for Mobile IPv4 conformed to the network access control architecture described in Chapter 4. The wireless terminal supports the Supplicant, the foreign agent supports the Authenticator, and the home agent supports the Account Authority, although the home agent typically passes AAA traffic to an AAA server rather than handling the traffic itself. Alternatively, the foreign agent can send the wireless terminal's authentication information directly to an AAA server, and the home agent can verify the authentication with an AAA server when processing a registration request. Table 6.4 contains a mapping between the network access control architecture functional elements and interfaces described in Chapter 4 and the security protocols defined for Mobile IPv4.

The Mobile IPv4 authentication protocol on the N1, N2, and N3 interfaces is primarily described in RFC 4721. The protocol in RFC 4721 is an extension to Mobile IPv4 registration that allows authentication information to be sent between the various entities. RFC 2794 describes a protocol for identity management that can also be used on N1, N2, and N3. The protocol in RFC 2794 is an extension to the Mobile IPv4 registration signaling that allows a wireless terminal to send an NAI to the foreign agent and home agent. RFC 3957 describes a key management technique and protocol to allow the

Table 6.5 Mapping of interfaces to protocols for Mobile IPv4 without Foreign Agent and Mobile IPv6

	Protocols on interfaces			
	TH1	TH2	TC1	TC2
Mobile IPv4 without Foreign Agent	– RFC 3957 – RFC 2794	– RFC 4721	<None>	<None>
Mobile IPv6	– IKEv1 (RFC 2409) – IKEv2 (RFC 4306)	– IPSec Encapsulating Security Payload (ESP) (RFC 4303)	– Return Routability Protocol (RFC 3775) – Manual Configuration (RFC 4449) – Enhanced Route Optimization (RFC 4866)	– Binding Authorization Data Mobility Header Option

wireless terminal, foreign agent, and home agent to set up security associations and derive shared keys. If RFC 4721 and RFC 2792 are used on the foreign agent to home agent interface, the authentication and NAI options are included in the Mobile IPv4 registration signaling. A foreign agent can also use Radius to send the information in these options directly to an AAA server. If Radius is used, vendor-specific Radius attribute/value pairs convey the information between the foreign agent and the AAA server, and between the AAA server and the home agent. The N4 interface is not defined for Mobile IPv4, since Mobile IPv4 does not specify security for data traffic between the wireless terminal and foreign agent. There are also some RFCs not cited here which describe proprietary extensions for Mobile IPv4 security used only in certain vendors products or particular deployments.

If a foreign agent is not present in the access network, however, the Mobile IPv4 architecture conforms more closely to the IP mobility management architecture described in this chapter. Table 6.5 provides a rough mapping between the functional elements and interfaces in the IP mobility architecture and the Mobile IP4 security protocols when a foreign agent is not present. The mapping between the IP mobility management architecture and the Mobile IPv6 security protocols is also presented in the table. In both cases, the wireless terminal supports the Mobile Node, the home agent supports the Home Agent, and the correspondent node supports the Correspondent Node. As should be evident from the tables and this discussion, the Mobile IPv4 protocol, especially the security protocols for Mobile IPv4, were not designed with any specific architecture in mind. Instead, options and extensions were grafted onto the original protocol as problems arose in various deployments. The resulting collection of protocols, options, extensions, and vendor-specific customizations is difficult to characterize in any consistent, organized fashion. In contrast, Mobile IPv6 was designed with a specific architecture in mind, and the security architecture therefore is cleanly separable into specific interfaces with protocols regardless of deployment circumstances. The remaining sections in this chapter discuss the protocols involved in Mobile IPv6 security.

6.3.6   Mobile IPv6 interfaces and protocols

The Mobile IPv6 protocol in RFC 3775 introduces the Binding Update/Binding Acknowledgement (or Binding Error) messages to perform binding management. These messages are sent between the wireless terminal and both the home agent – for home agent binding management – and the correspondent node – for route optimization. The binding management protocol is implemented as an IPv6 header, the Mobility Header, and a collection of Mobility Header Options. RFC 3775 also introduces another mobility-related header, the Type 2 Routing Header and two options for another header, the Home Address Option and the Alternate Care-of Address Option. The security implications of the Type 2 Routing Header, Home Address Option, and Alternate Care-of Address option are discussed in more detail later in the chapter.

Rather than develop a new protocol for the security of binding update on the TH1 and TH2 interfaces, Mobile IPv6 uses existing IETF standardized protocols. On the TH1 interface, the Internet Key Exchange (IKE) protocol is used for establishing a security association between the Wireless terminal and home agent. IKE is a standard for identity authentication and key provisioning between any two IP nodes (both IPv4 and IPv6) at the internetworking level (IP layer) of the stack. On the TH2 interface, IP Security (IPsec) is used for actually protecting signaling and data traffic between the home agent and wireless terminal. IPsec has two protocols, Authentication Header (AH) and Encapsulating Security Payload (ESP). Both provide data origin authentication and replay protection, ESP also provides confidentiality protection. Because AH is rarely used, ESP is recommended for Mobile IPv6 wireless terminal to home agent security. IKE is described in RFC 2409 (RFC 2409, 1998) (for IKEv1) and RFC 4306 (RFC 4306, 2005) (for IKEv2), ESP is described in RFC 4303 (RFC 4303, 2005), and AH is described in RFC 4302 (RFC 4302, 2005). RFC 4301 (RFC 4301, 2005) provides an overview of the security architecture for the IP networking layer, As discussed in Chapter 4, IKE and IPsec are widely deployed for virtual private networks, and they are recommended over public access hotspot networks where the network itself provides no data origin authentication or confidentiality protection. In the next sections, we describe IPsec, ESP, and IKEv2 and how they are used in Mobile IPv6. We use IKEv2 as an example because it is considerably simpler than IKEv1 and therefore easier to understand. The simplicity is also expected to result in wider deployment.

Rather than using IKE and IPsec, the Mobile IPv6 specification and a few additional RFCs define new security protocols on the TC1 and TC2 interfaces. Route optimization is a fundamentally new capability in the IP network architecture. It cannot require identity verification like the home agent/wireless terminal verification because the Internet does not support a generalized method of verifying identity between two random nodes. In the home agent/wireless terminal case, a preexisting business or other relationship ensures that the two sides can easily verify each others' identity.

6.4   The IP Security (IPsec) protocol

The IP Security protocol suite provides security at the IP (networking) layer (or Layer 3).
The IP Security protocol suite consists of two protocols:

- the Internet Key Exchange (IKE) for authenticating node identity and establishing a security association containing shared state, including key provisioning, between two nodes wanting to use IP Security;
- the IP Security (IPsec) protocol itself for protecting IP packets with data origin authentication, confidentiality protection, and anti-replay protection on traffic between the two nodes.

IKE and IPsec were designed to be used between any two nodes in the Internet that want to protect traffic at the IP networking layer, for both IPv4 and IPv6. For Mobile IPv6, IKE and IPsec protect binding update and home link configuration information exchange between the wireless terminal and home network.

The following sections present overviews of the basic IP security architecture, the design of IKEv2 which is the latest version of IKE, and IPsec Encapsulating Security Payload (ESP), and how they are used in Mobile IPv6. For IKE, the emphasis is on understanding the protocol semantics rather than the details of the message syntax. For IPsec, the protocol semantics are simple, and so more emphasis is placed on the message syntax. In both cases, consult the relevant Internet RFCs for complete details, particularly if implementation is intended.

6.4.1   The IPsec architecture

IPsec includes two separate security services for the network layer:

- Authentication Header (AH) provides data origin authentication on the entire packet, including the IP header and certain options, depending on the IP version.
- Encapsulating Security Payload (ESP) provides data origin authentication and/or confidentiality protection on the contents of the packet, exclusive of the header and certain options.

In addition to these services, anti-replay protection is provided if dynamic key provisioning is used. In Mobile IPv6, both services are supported, but ESP is expected to be more widely used because it supports both data origin authentication and confidentiality protection in a single protocol and both are needed.

These services are available in two different modes:

- Transport mode provides end-to-end services for traffic sent from one node and received and processed by another. The IPsec header appears between the IP header and certain options and the next layer header, which is typically the header for a transport protocol such as TCP or UDP.
- Tunnel mode provides a secure tunnel for traffic flowing through a gateway node. The IPsec header in this case is included after an outer IP header and options that specifies the tunnel end point at the gateway node where the packet is decapsulated for further delivery. An inner IP header specifies the final destination of the packet, followed by the packet contents.

The Mobile IPv6 wireless terminal to home agent connection utilizes both transport mode and tunnel mode.

Within an IP node, IPsec is used to establish protection on particular IP interfaces. IPsec creates a boundary between protected and unprotected networking interfaces on the node. Any traffic that moves across the boundary is subject to access control and security services processing by one of the two security services provided by IPsec. There are three possible actions that IPsec can take as a packet traverses the security boundary:

- Discard the packet if the packet matches an IPsec policy database template that indicates it should be discarded or if the packet does not match any IPsec policy database template.
- Process the packet if the packet matches an IPsec policy database template that indicates it should be processed.
- Bypass IPsec if the packet matches an IPsec policy database template that indicates it should bypass IPsec processing, because the packet is from a privileged source (typically only applied to IKE packets).

The overall effect is somewhat like a firewall, except the security services supported are more sophisticated because the processing may involve cryptographic operations in addition to simply keeping or dropping the packet.

IPsec requires that two nodes that are engaged in mutual security operations share a security association (SA). An IPsec SA is a collection of state that applies to the unidirectional traffic flow between nodes. Most protocol transactions consist of bidirectional traffic flows, so there are typically two SAs between two nodes in most uses of IPsec, one for each direction. The opposite side has SAs that point in the opposite directions.

An IPsec SA consists of the following state shared between the two sides:

- the security services (data origin authentication, confidentiality and anti-replay protection) that are provided for processing the packets between the two nodes and which protocols (AH or ESP) and modes (transport or tunnel) are used;
- the cryptographic algorithms used to provide these services;
- the corresponding keys for the cryptographic algorithms.

SAs can be set up by any means but RFC 4301, which describes the IPsec architecture, discusses only two: manual provisioning and IKE for dynamic provisioning. Because manual provisioning does not scale well, IKE is commonly used to provide scalability when IPsec is needed. Both manual provisioning and IKE are supported for Mobile IPv6.

There are three databases specified by the IPsec architecture and associated with the IPsec implementation on a particular node. The databases and their contents are:

- The Security Policy Database (SPD) which contains security policy templates matching against traffic crossing the IPsec boundary on the node.
- The Security Association Database (SAD) which contains the shared state for specific security associations that have been either assigned manually or dynamically negotiated between two nodes.

- The Peer Authorization Database (PAD) which links a particular policy template in the SPD with an SA management protocol such as IKE.

The subsections below provide more detail on the three databases.

Security Policy Database

The templates in the SPD are used to identify which packets to process with IPsec. Each selector has one of three types:

- SPD Secure (SPD-S) selectors specify that matching traffic is subject to IPsec processing.
- SPD Outbound (SPD-O) selectors are for traffic exiting the node and specify that matching traffic either bypasses IPsec processing or is discarded.
- SPD Inbound (SPD-I) selectors are for traffic entering the node and specify that matching traffic either bypasses IPsec processing or is discarded.

Selectors within the database are applied to packets, and if a packet matches one of the selectors, the appropriate action is taken depending on the selector type. If the packet does not match any selector, it is discarded.

A selector contains a template that is matched against packets. The content of the template can contain any combination of the following items:

- The special values ANY and OPAQUE indicate cases where the field value can have an arbitrary value or need not be present, for example if the packet is encrypted.
- Unicast IP addresses in various combinations may appear. Single addresses, lists of addresses, or ranges of addresses can be included in the selector, with both global and link local scope.
- Next layer protocol types, obtained from the IPv6 "Next Header" field specify the protocol at the next layer in the packet. The selector value is an individual protocol number, ANY or, in the case of IPv6 only, OPAQUE.
- If the next layer protocol is a transport layer using ports (TCP, UDP, SCTP, etc.), then the selector can include a local or remote port template, with a single port, port list, port range or the special values ANY and OPAQUE.
- If the next layer protocol does not use ports (ICMP and the Mobility Header in IPv6 which is used for Mobile IPv6), the template is a message type identifier. For ICMP, the identifier is the message type and code, while for the Mobility Header the template is the 8-bit mobility message type. For ICMP, the template can contain a range of types and codes, and the special value ANY.
- Finally, the selector template can contain a name. Unlike the other template values, the name is not matched against a field in the packet but is used during SA negotiation with IKE.

Security Association Database

The SAD contains the shared security state for one direction (i.e. either incoming or outgoing) indicated by the matching template in the SPD. The two database entries are tied together by the Security Parameters Index (SPI), which is a 32-bit security association identifier included by all packets protected with IPsec in clear text. For traffic entering the node, the SPI identifies which SAD entry to use when processing a packet matching an SPD-S template. The IPsec implementation processing incoming traffic can bypass the SPD and go directly to the SAD because the SAD contains the cached SPD template. For traffic exiting the node, the IPsec implementation matches the packet against the SPD selector templates to identify which SAD entry (if any) should be applied to the packet for IPsec processing, and uses the SPI to retrieve the corresponding shared security state in the SA. When the outgoing packet is released, the IPsec header contains the SPI so that the node on the other end of the connection can identify which SA to use when processing the packet.

The SAD contains a comprehensive list of fields covering IPsec processing, in addition to cached template values from the SPD. The information in these fields covers:

- The 32-bit SPI.
- A 64-bit sequence number and sequence counter overflow flag indicating what to do (i.e. rollover or restart) if the sequence number overflows.
- If the SA is dynamically negotiated, an anti-replay field containing a 64-bit counter and bit map.
- The cryptographic algorithm, key, other cryptographic parameters, and service type of the SA. The four possible service types are AH, ESP for data origin authentication, ESP for confidentiality protection, and ESP for both.
- The lifetime of the SA.
- The mode of the SA, i.e. transport mode or tunnel mode. If the SA is tunnel mode, the source and destination IP addresses of the endpoints (which must both be of the same IP version) are also included.

Peer Authentication Database

The PAD links the SPD template entry to a protocol, such as IKE, that dynamically negotiates the security association and creates the SAD entry. PAD entries specify which nodes or groups of nodes are authorized to communicate with the node. The protocol (typically IKE) and authentication method is used to authenticate other nodes. The authentication data is for conducting an authentication. The PAD also includes information to constrain the identities that other nodes are allowed to assert during the authentication transaction, and may include information on the location of security gateways for nodes where a security gateway acts as an intermediary. The PAD supports a variety of name formats for identity assertion, including NAI-like names, DNS names, and IP addresses.

6.4.2 IKE

The main protocol for two nodes to dynamically negotiate an IPsec security association is the Internet Key Exchange (IKE) protocol. IKE allows any two nodes on the Internet to mutually authenticate their identities and set up IPsec SAs. The original version of IKE, now called IKEv1, is defined in RFC 2409. It is a very complicated protocol, involving the amalgamation of two different key exchange protocols together with a "framework" protocol for key exchange that did not specify a specific key exchange protocol. As a consequence, IKEv1 never achieved wide implementation and deployment success. It has been deployed primarily in Virtual Private Network (VPN) systems, where it achieved moderate success, but designers of other protocols and systems failed to pick it up. In addition, IKEv1 lacked some fundamental functionality, such as AAA-based identity authentication, that is important for backward compatibility. This functionality was added as vendor extensions but interoperability was not assured because the functionality was not defined in the standard.

As a consequence, the IETF redesigned IKE from the bottom up and in 2005 issued RFC 4306 for IKEv2. The resulting protocol is considerably simpler. In many cases, it is possible to complete peer identity authentication, establish an IKE SA between the peers, and establish a simple SA for IPsec with four messages, a radical reduction over the complex IKEv1 protocol. The expectation in the IETF security community is that IKEv2 would be adopted more widely by other protocols and systems, and, indeed, this seems to be happening. While there are still a few minor key provisioning scenarios not covered by IKEv2, in many cases where establishing an SA for IP level security is required, IKEv2 can conveniently and securely provide it.

IKEv2 Protocol

IKEv2 is a request/response protocol in which the initiator sends a request to the responder and the responder replies. An IKEv2 transaction consists of two required and one optional request/response exchange:

1. The IKE_SA_INIT exchange in which the two sides negotiate cryptographic algorithms, exchange nonces for child SA key generation, and do a Diffie–Hellman exchange to establish a shared key for the following IKE exchanges.
2. The IKE_AUTH exchange in which the previous messages are authenticated, the identities of both sides are authenticated, and a simple IPsec SA, called a child SA, is established.
3. The optional CREATE_CHILD_SA exchange in which additional and/or more complex child SAs are established. This exchange is unnecessary if the simple SA established by the IKE_AUTH exchange is sufficient.

The first two exchanges establish an SA for the IKE transaction itself. They can also be used to establish a simple child SA, if only a single child SA is needed. The third exchange is required only for more complex child SAs or additional child SAs. The IKE_SA_INIT and IKE_AUTH exchanges must be completed in order before any CREATE_CHILD_SA exchanges can complete. IKEv2 also supports an INFORMATIONAL exchange. The INFORMATIONAL exchange performs housekeeping functions such as deleting an SA, reporting error conditions, or checking whether a peer is still alive. INFORMATIONAL exchanges, like the CREATE_CHILD_SA exchange, run after the IKE SA has been established.

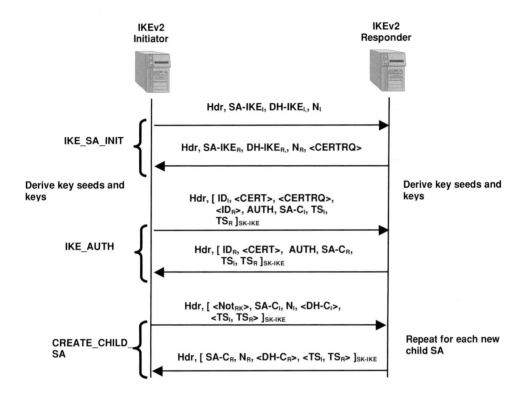

Figure 6.4 Basic IKEv2 exchange

Figure 6.4 illustrates the basic IKEv2 IKE_SA_INIT, IKE_AUTH, and CRE-ATE_CHILD_SA exchanges. In the figure, values in angle brackets ($<\,>$) are optional and square brackets ([]) indicate that the contained values are protected with authentication and encryption.

Each message consists of an IKE protocol header (Hdr in the figure) followed by a sequence of payload values. The IKE_SA_INIT exchange starts with a message sent by the initiator to the responder with the following payloads:

- A security association descriptor (SA-IKE$_I$) describing the cryptographic algorithms the initiator supports for the IKE SA.
- The initiator's Diffie–Hellman values (DH-IKE$_I$) for provisioning a shared key.
- The initiator's pseudo random nonce (N$_I$) for key generation.

The responder replies with a message containing the following payloads:

- A security association description (SA-IKE$_R$) selecting one of the cryptographic algorithms from the initiator's collection for the IKE SA.
- The responder's Diffie–Hellman values (DH-IKE$_R$) for provisioning the IKE SA shared key.
- The responder's pseudo random nonce (N$_R$) for key generation.

- An optional certificate request (CERTRQ) containing trust anchor certificates if the responder would like to authenticate the initiator using an X.500 certificate.

At this point, both sides have enough information to generate a seed key, utilizing the Diffie–Hellman parameters and the nonces, from which all shared keys for the IKE SA are derived. The IKE SA authentication and encryption shared keys derived from the seed key are applied to all payload data in further exchanges (indicated in the figure by: []$_{SK-IKE}$).

The IKE_AUTH exchange starts with the initiator sending a message to the responder containing the following payloads:

- The initiator's IKE identity (ID$_I$).
- An optional certificate (CERT) if the responder asked for one allowing the responder to authenticate the initiator's identity.
- An optional certificate request (CERTRQ) containing trust anchor certificates if the initiator would like to authenticate the responder using an X.500 certificate.
- An optional IKE identity that the initiator expects in the responder's reply (ID$_R$).
- An authenticator used to prove the initiator's identity (AUTH), i.e. a digital signature if a public key algorithm is used or a MAC if a preshared key and secret key algorithm is used.
- A security association description (SA-C$_I$) indicating the cryptographic algorithms the initiator supports and the security service requested for the first child IPsec SA.
- A pair of traffic selectors for the initiator and responder (TS$_I$, TS$_R$) for the first child SA. These traffic selectors are taken from the initiator's SPD and indicate the SPD entries that should match the child SA traffic for the initiator and responder.

The responder replies with a message containing the same payloads as in the IKE_AUTH initiator message, except they apply to the responder. As in the initiator's message, the CERT payload is included only if the initiator asked for one.

At this point, the IKE transaction has been authenticated and a single child SA has been established. If no other child SAs are required, the IKE transaction terminates here. If, however, additional child SAs are required, the transaction moves to the CRE-ATE_CHILD_SA exchange.

The initiator starts the CREATE_CHILD_SA exchange with a message containing the following payloads:

- If the CREATE_CHILD_SA exchange is rekeying an existing SA, then the leading payload is a NOTIFY type (Not$_{RK}$) with the SPI being rekeyed. If the SA is new, this payload is omitted.
- A security association description (SA-C$_I$) describing the cryptographic algorithms the initiator supports and the security service sought for the child SA.
- The initiator nonce (N$_I$).
- An optional set of Diffie–Hellman parameters for provisioning the child SA shared key (DH-C$_I$), if the initiator wants to derive the child SA keys from a new root.
- Unless the CREATE_CHILD_SA is for rekeying, a pair of traffic selectors for the initiator and responder (TS$_I$, TS$_R$).

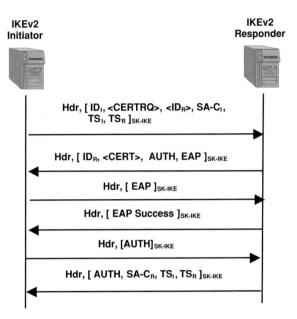

IKEv2
Initiator

IKEv2
Responder

Hdr, [ID$_I$, <CERTRQ>, <ID$_R$>, SA-C$_I$,
TS$_I$, TS$_R$]$_{SK-IKE}$

Hdr, [ID$_R$, <CERT>, AUTH, EAP]$_{SK-IKE}$

Hdr, [EAP]$_{SK-IKE}$

Hdr, [EAP Success]$_{SK-IKE}$

Hdr, [AUTH]$_{SK-IKE}$

Hdr, [AUTH, SA-C$_R$, TS$_I$, TS$_R$]$_{SK-IKE}$

Figure 6.5 IKE_AUTH with EAP for IKE identity authentication

The responder replies with exactly the same set of values, except they apply to the responder and the NOTIFY payload is not present even for rekeying.

EAP for IKE_AUTH

One feature of IKEv2 that is important for Mobile IPv6 is the use of Extensible Authentication Protocol (EAP) instead of preshared keys or certificates for authenticating the IKE identity during the IKE_AUTH exchange. Many mobile network service providers have AAA databases for network access authentication, and they would like to leverage those databases to authenticate clients for IP mobility service as well. Using AAA avoids having to provision the clients with a preshared key or certificate. Figure 6.5 shows a minimal IKE_AUTH exchange when EAP is used for the IKE identity authentication. Note that EAP only applies to the authentication of the initiator's identity by the responder, the responder still includes an AUTH payload, authenticated with the public key in the certificate, in its second response.

The initiator indicates a preference to use EAP by including an IKE identity but leaving out the AUTH payload from the first message in the IKE_AUTH exchange. If the responder can accommodate using EAP, it includes an EAP message in the response to initiate the EAP exchange. The responder child SA and traffic selectors are not sent until the EAP exchange has completed and the initiator's identity has been authenticated. The exchange can then continue for a number of messages, since the initiator and responder might need to negotiate the EAP method to use among other tasks. If the EAP exchange successfully completes, the responder sends an EAP Success message. The initiator then sends a message with its identity authentication AUTH payload. The responder replies with its identity authentication AUTH payload and the child SA and traffic selectors for the child SA. If the EAP identity authentication fails, the responder sends EAP Failure.

If the EAP method used supports shared key provisioning, the key used for calculating the final AUTH by both the initiator and the responder is the EAP Master Session Key (MSK). This shared key is not used for any other purpose. If the EAP method does not establish a shared key, the AUTH payload is created using a new set of authentication keys generated from the IKE seed key root on the initiator and responder. In general, an EAP method that does establish a shared key should be used because there are certain man-in-the-middle attacks that can be mounted against any EAP method if the method does not establish a secure tunnel to the AAA server.

6.4.3   IPsec Encapsulating Security Payload

IPsec Encapsulating Security Payload (ESP) defines a packet format and mechanism whereby data origin authentication (which includes integrity protection), confidentiality protection, and anti-replay protection are provided on a traffic flow between two nodes. An encapsulator surrounds the packet contents to be protected, for transport mode, or the entire packet, for tunnel mode. The ESP encapsulator starts immediately after the IPv4 header, or after any IPv6 header options which are processed in transit (e.g. hop-by-hop, routing, fragmentation, and some destination options). An ESP protected packet is indicated by the identifier 50 in the Protocol field (for IPv4) or Next Header Type field (for IPv6) of the header or option immediately preceding the start of the ESP encapsulator.

Figure 6.6 illustrates the format of the ESP encapsulation. The fields have the following definitions:

- Security Parameters Index (SPI) – a 32-bit field indicating the security association to use during ESP processing. The SPI is established between the two nodes when the security association is negotiated. The receiver uses the SPI to look up the security association in the SAD.
- Sequence Number – a 32-bit field used for anti-replay protection. Both sides initiate the sequence number to zero when the security association is established and increment it for each packet sent under protection of the security association. If anti-replay protection is not enabled, the receiving node ignores the field. IPsec also supports an optional 64-bit sequence number to reduce the incidence of field overflows, but only the low-order bits are inserted in the field. The high-order bits are included in the authenticator calculation.
- Encapsulated Payload – This variable-length field contains the data protected by the ESP security association. If the security association is in transport mode, this field contains bytes from the transport layer on up. If the security association is in tunnel mode, this field contains an entire IP packet, header and all.
- Padding – an optional padding field contains between 0 and 255 bytes filling out the Encapsulated Payload field to right align the Pad Length and Next Header fields within a four byte word. In many cases, the content of the padding bytes is specified by the

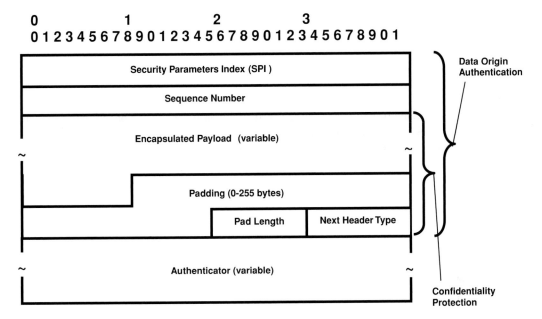

Figure 6.6 Format of IPsec ESP encapsulated packet

encryption algorithm, but if that is not the case, the bytes are initialized with a series of one byte values, starting at 1 and increasing sequentially.

- Pad Length – the number of bytes in the Padding field.
- Next Header Type – the type of the next header or zero if there is none.
- Authenticator – If data origin authentication (which includes integrity checking) is provided by the security association, either alone or in combined mode with confidentiality protection, a variable-length authenticator may be included at the end before the next header if any. Some combined mode algorithms may include the authenticator in the Encapsulated Payload, in which case the Authenticator field is omitted. The actual contents and size of the Authenticator field depend on the authentication algorithm. The authenticator is computed over the contents of the ESP header (the SPI and Sequence Number fields), Encapsulated Payload, and ESP trailer (Padding, Pad Length, and Next Header Type fields).

In addition to these standard fields, an ESP protected packet may contain some optional fields within the Encapsulated Payload field depending on the nature of the cryptographic algorithms in the security association. They are:

- Initialization Vector – Some encryption algorithms require cryptographic synchronization data in the form of an initialization vector. If an initialization vector is required, it must be placed at the beginning of the Encapsulated Payload field in unencrypted form.
- Combined Mode Authenticator – In combined mode, the authenticator may be included in the Encapsulated Payload field and the Authenticator field may be omitted.

- Traffic Flow Confidentiality – RFC 4303 includes some advice about how to achieve traffic flow confidentiality to foil traffic analysis. If traffic flow confidentiality is enabled, additional padding bytes may be included in the Encapsulated Payload field beyond the Padding field bytes.

If any of these optional fields are included within the Encapsulated Payload field, their format and processing is documented in the specification of how the cryptographic algorithms are integrated with IPsec.

Although IPsec supports three combinations of security services (confidentiality alone, data origin authentication alone, combined mode), confidentiality protection alone only defends against passive attacks. Anti-replay protection is also provided only if the security association supports data origin authentication. Consequently, combined mode is recommended if confidentiality protection is required. In combined mode, outgoing packet processing calculates the authenticator first, and then encrypts the payload with or without the authenticator depending on the algorithm.

6.4.4   How Mobile IPv6 uses IKE and IPsec

Mobile IPv6 uses IPsec between the wireless terminal and the home agent in both transport and tunnel mode. The primary service used is data origin authentication but confidentiality protection is also required for some control messages and can be optionally used for both control and user traffic if the wireless terminal wants to obscure the contents. The security association between the home agent and the wireless terminal can be manually configured, but IKE is preferentially used for better scalability. The IKE SA can be authenticated by preshared key, certificate, or with EAP, but since most mobile network service providers have existing AAA databases for network access authentication, EAP may be the easiest option to deploy.

The following IPsec SAs are required for Mobile IPv6 traffic between the wireless terminal and home agent:

- An ESP transport mode SA for Binding Update and Binding Acknowledgement with security services for data origin authentication and optional anti-replay protection. Anti-replay protection is optional because the Binding Update message itself uses a sequence number to protect against replay attacks.
- An ESP tunnel mode SA for the return routability messages Home Agent Test Initiate and Home Agent Test with security services for data origin authentication, confidentiality protection, and anti-replay protection (see next section for a discussion of return routability).
- An ESP transport mode SA for the home subnet configuration messages Home Subnet Prefix Solicitation and Home Subnet Prefix Advertisement with security services for data origin authentication and anti-replay protection.

The transport mode SAs can optionally be replaced by tunnel mode SAs and confidentiality protection can be added if the privacy of the signaling connection between the

Table 6.6 Peer Authorization Database entries for Mobile IPv6

	Filter rules	Actions
Wireless terminal	remote identity == home agent identity	Authenticate and authorize CHILD_SA for home agent remote address
Home agent	remote identity == wireless terminal identity	Authenticate and authorize CHILD_SAs for wireless terminal home address

Table 6.7 Security Policy Database for Binding Update/Binding Acknowledgement

	Filter rules	Actions
Wireless terminal	(local address == home address) && (remote address == home agent) && (protocol == Mobility Header) && (local message type == Binding Update) && (remote message type == Binding Acknowledgement)	Use ESP transport mode, initiate using wireless terminal user identity to address home agent
Home agent	(local address == home agent) && (remote address == home address) && (protocol == Mobility Header) && (local message type == Binding Acknowledgement) && (remote message type == Binding Update)	Use SA for ESP in transport mode

Table 6.8 Security Policy Database for Home Network Prefix Discovery

	Filter rules	Actions
Wireless terminal	(local address == home address) && (remote address == home agent) && (protocol == ICMPv6) && (local message type == Home Network Prefix Solicitation) && (remote message type == Home Network Prefix Advertisement)	Use ESP transport mode, initiate using wireless terminal user identity to address home agent
Home agent	(local address == home agent) && (remote address == home address) && (protocol == ICMPv6) && (local message type == Home Network Prefix Advertisement) && (remote message type == Home Network Prefix Solicitation)	Use SA for ESP in transport mode

wireless terminal and home agent is of concern. In addition, for maximum confidentiality, an ESP tunnel mode SA with data origin authentication, anti-replay protection, and confidentiality protection can be maintained for all data traffic tunneled between the wireless terminal and home agent. When tunnel mode SAs are used, the signaling and/or data traffic is completely encrypted so an eavesdropper knows only what is in the IPv6 header.

The configuration of the IKEv2 Peer Authorization Database and Security Policy Database for Mobile IPv6 is shown in Tables 6.6 through 6.10. Similar database entries

Table 6.9 Security Policy Database for return routability

	Filter rules	Actions
Wireless terminal	(local address == home address) && (remote address == any address) && (protocol == Mobility Header) && (local message type == Home Address Test Initiate) && (remote message type == Home Address Test)	Use ESP tunnel mode, initiate using wireless terminal user identity to address home agent
Home agent	(local address == any) && (remote address == home address) && (protocol == Mobility Header) && (local message type == Home Address Test) && (remote message type == Home Address Test Initiate)	Use SA for ESP in tunnel mode

Table 6.10 Security Policy Database for ESP protected traffic tunnel

	Filter rules	Actions
Wireless terminal	(interface == IPV6 tunnel to home agent) && (source == home address) && (destination == any) && (protocol == any)	Use ESP tunnel mode, initiate using wireless terminaluser identity to address home agent
Home agent	(interface == IPV6 tunnel to home address) && (source == any) && (destination == home address) && (protocol == any)	Use SA for ESP in tunnel mode

are necessary if manual SA configuration is used instead of IKE. The IPsec implementation matches incoming or outgoing packets against the filter rules, and applies the actions if the rules match. In addition, if the home address is dynamically configured (see below), the home agent must maintain an SPD template and create new SPD entries when the home address becomes known. Similarly, the PAD entries must be dynamically created if a home address is dynamically assigned.

The IKEv2 exchange used to dynamically provision SAs on the wireless terminal and home agent is the same as in Figure 6.4 if a preshared key or bidirectional certificate exchange is used for authentication or in Figure 6.5 if EAP is used. In addition to the standard IKE configuration payloads, Mobile IPv6 allows the wireless terminal to configure a dynamically assigned home address. The wireless terminal requests a dynamically assigned home address by including a configuration payload request (CFG_REQUEST) of type INTERNAL_IP6_ADDRESS into the IKE_AUTH exchange. If the wireless terminal has no preference for an address, it sets the suggested address field to zero, otherwise, a suggested address can be included. The home agent replies with a configuration payload reply (CFG_REPLY) of type INTERNAL_IP6_ADDRESS with the address and the length of the subnet prefix. The home agent can obtain the address either from the suggested address, from a DHCP request, or by any other means. The lifetime of the address is the same as the SA's, and is extended if the SA is rekeyed. If the home agent cannot allocate an address for the wireless terminal, it replies with a Notify payload having an INTERNAL_ADDRESS_FAILURE message, and the wireless terminal must restart the IKE exchange or switch to a different home agent.

Mobile IPv6 imposes the following specific constraints on the values incorporated into the IKE payloads in Figures 6.4 and 6.5:

- The wireless terminal uses its care-of address only as the source address on the IKE packets. The initiator address used in the traffic selectors (TS_I and TS_R) to specify the filters used in the SA must be the home address. This allows the SPD entries to survive a care-of address change.
- The wireless terminal's IKE identity should never be the care-of address, and the home address also cannot be used if it is dynamically assigned. In practice, the home address should not be used even if it is statically assigned unless there is no other choice. The identity is typically an NAI-like identifier or, if a certificate is used for authentication, the identity can be a fully qualified domain name. In addition, if EAP is used for authorization of the wireless terminal, the wireless terminal may use a different identity during the EAP transaction than its IKE identity. The IKE identity may be the NAI while the actual EAP identity may be something more, such as a user name/password.

There are also a few constraints on the format and processing of the actual IPsec-protected signaling and data packets that are required in order for IPsec processing to succeed:

- Because the IPsec processing uses the home address to match the filter rules and not the care-of address, the home address must be visible outside the ESP-protected portion of the packet. A packet sent from the wireless terminal has the care-of address as the source address in the IPv6 header, and similarly a packet sent from the home agent has the care-of address as the destination address, so the header addresses cannot be used for the IPsec processing. Instead, the wireless terminal includes a new IPv6 destination option, the Home Address Option, with the home address. The home agent includes a new type of routing header, the Type 2 Routing Header. These options are outside the IPsec protected part of the packet and contain the home address, which can then be matched against the filter rules.
- Because ESP does not protect the packet header (nor the Home Address Destination Option or Type 2 Routing Header), Binding Updates protected by ESP from the wireless terminal must include an Alternate Care-of Address Mobility Option within the ESP-protected part of the packet. The home agent uses this address to determine the new care-of address and not the source address on the packet.
- One particularly tricky issue is how to handle care-of address changes. In particular, the ESP tunnels used for protection of return routability signaling and data packets have the care-of address as their endpoint, and the SA entry in the SAD must be modified

with the new care-of address. In addition, the IKE SA itself needs to have the care-of address updated. RFC 3775 recommends that the home agent implement a private API used only by the Mobile IPv6 implementation to allow these changes. The API must never be exposed to nonprivileged clients, since such use could open up security vulnerabilities. The Mobile IPv6 Binding Update and Binding Acknowledgement messages have a Key Management Capability flag for the home agent to indicate whether this API is available or not. If it is not available, the wireless terminal must re-run IKE on every movement to a new access router, to re-do key provisioning.

6.5   Return routability

The protocol used for key provisioning and security on the TC1 and TC2 interfaces is the return routability protocol. While IKE and IPsec could have been used on these interfaces as well, authentication poses a problem. Route optimization can occur between the wireless terminal and any host on the Internet. The Internet lacks a comprehensive authentication infrastructure that covers all nodes, such as a global PKI or globally connected AAA database. In the absence of such an infrastructure, it is possible that the IKE authentication step or an AAA authentication between the wireless terminal and correspondent node may fail due to lack of interoperable authentication. Consequently, a key provisioning protocol was developed that does not depend on any shared trust reference between the wireless terminal and the correspondent node. The protocol ensures that the wireless terminal is, in fact, located at the care-of address and home address it claims to be located at when a key is provisioned, and that a Binding Update message issued by a wireless terminal having a particular care-of address did issue from the same wireless terminal with which the key was provisioned. This assurance derives from the basic nature of IP routing – that an uncompromised routing infrastructure always delivers packets to the right destination address – and not from any cryptographic property. The security is considerably weaker than the security on the wireless terminal/home agent interface, but is the best possible security for two random nodes given the lack of pervasive, standardized identity management on the Internet.

Figure 6.7 illustrates the return routability protocol. The protocol conducts a key provisioning from the correspondent node to the wireless terminal across two paths, one through the home agent and one directly between the wireless terminal and correspondent node. The protocol consists of four messages: Home Address Test Initialize (HoTI), Care-of Address Test Initialize (CoTI), Home Address Test (HoT), and Care-of Address Test (CoT). The protocol is run after the wireless terminal has sent a Binding Update to the home agent to change the home address binding to the new care-of address.

The wireless terminal reverse tunnels a HoTI message through the home agent bound for the correspondent node, and, at the same time, sends a CoTI message directly to the correspondent node using the care-of address directly as the source address. Both messages contain 64-bit randomly generated cookie values (C_H for the home init cookie and C_C for the care-of init cookie). The cookies are also included in the replies (HoT

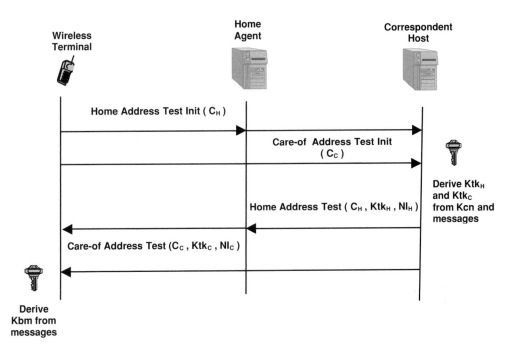

Figure 6.7 Return routability protocol

and CoT) to ensure that the right correspondent generated the reply, and to allow the replies to be matched with the requests.

The correspondent node maintains a 20-byte secret key for return routability key generation, Kcn, that is randomly generated. This key is not shared with any other entity, and it can be changed at any time. When the correspondent node receives a HoTI and CoTI message, it generates two keygen tokens, Ktk_H for the HoT and Ktk_C for the CoT:

$$Ktk_H = FIRST\,(64, HMAC\_SHA1(Kcn, (home\;address|N_H|\,0)))$$

$$Ktk_C = FIRST\,(64, HMAC\_SHA1(Kcn, (care\text{-}of\;address|N_C|\,1)))$$

Here, FIRST (64,.) indicates that the first 64 bits are extracted and HMAC_SHA1 applies the SHA1 message digest followed by the HMAC keyed hash to the concatenated values in the contents, using Kcn as the key. For Ktk_H, the digested value includes the home address, a home address test nonce, N_H, and a single byte, 0x00, concatenated on the end. For Ktk_C, the digested value includes the care-of address, the care-of address test nonce, N_C, and a single byte, 0x01 concatenated on the end. The last byte in each case allows the correspondent node to distinguish between the tokens. The nonces are randomly generated bit strings that are changed periodically. They can be any length but 64 bits is the recommended value. The correspondent node keeps track of nonces using indices, and sends the index values (NI_H and NI_C) to the wireless terminal in the HoT and CoT messages. The nonces are never communicated directly to the wireless terminal.

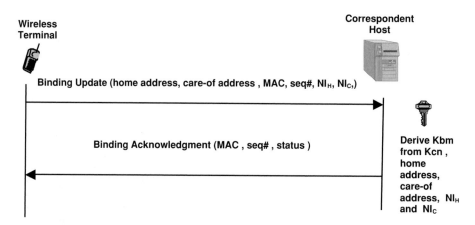

Figure 6.8 Binding Update protocol with authorization

After generating the key tokens, the correspondent node replies to the HoTI and CoTI messages with the HoT and CoT messages, as shown in the figure. The cookies from the request messages, and the respective key generation tokens and nonce indices are included in the HoT and CoT messages. When the wireless terminal has received both messages, it generates the binding management key, Kbm, using the following algorithm:

$$Kbm = SHA1(Ktk_H \mid Ktk_C)$$

The binding management key is used to generate a shared key MAC on the Binding Update message. Figure 6.8 illustrates the binding update protocol with authorization. The wireless terminal sends the Binding Update message (including the home address) in the Home Address Destination Option. The Binding Update message also includes the care-of address as the source address or in the Alternate Care-of Address Option. The MAC for authorization, a sequence number for anti-replay protection, and the two nonce indices complete the security protocol. The MAC is calculated using the following algorithm:

$$MAC = FIRST\,(96, HMAC\_SHA1(Kbm, (care\text{-of address} \mid$$
$$correspondent\ node\ address \mid Binding\ Update)))$$

The authorization MAC is included in the Binding Authorization Data Option.

The correspondent node uses the master secret key for binding key generation, Kcn, the home address, the care-of address, and the two nonces retrieved using the nonce indices to re-create Ktk_H and Ktk_C and thereby Kbm. Note that this procedure allows the correspondent node to utilize the nonces for multiple clients, thereby saving memory and avoiding a potential DoS attack vulnerability that would allow an attacker to launch a state depletion attack by sending repeated return routability messages. In addition, the correspondent node can change the nonces at any time, as long it maintains a record of the old nonces for outstanding binding management keys.

The correspondent node uses Kbm to verify the MAC. If the MAC verifies, the binding is changed and the correspondent node optionally replies with a Binding Acknowledgement, including a MAC there as well. The MAC on the Binding Acknowledgement is calculated as:

$$\text{MAC} = \text{FIRST}\,(96, \text{HMAC\_SHA1}(\text{Kbm}, (\text{care} - \text{of address}|$$
$$\text{correspondent node address}|\text{Binding Acknowledgement})))$$

The basis of the return routability protocol is the assumption that in the Internet without Mobile IP, the routing infrastructure is secure meaning packets have a high probability of reaching their destination without modification. Compromising a router is extremely hard since ISPs tend to defend them strongly and compromising routing through a man-in-the-middle attack is fairly rare and difficult to do, with one notable exception. That exception is on the last hop between the access router and an end host, where an attacker can interpose and portray itself as the access router.

Because the two halves of the binding management key are sent in the clear, return routability is not secure against attacks that can eavesdrop on traffic between the wireless terminal and both the home agent and the correspondent node at the same time. Since most routers within the network are well defended, the most likely place that such an attack could occur is on the last hop between the access router and the correspondent node or the wireless terminal; for example, on an 802.11 network with no over-the-air security between the host and the access point. The risk of compromise on the wireless terminal side is mitigated by using confidentiality protection on the tunnel between the wireless terminal and the home agent for the HoTI/HoT transaction. The HoTI/HoT transaction between the wireless terminal and the home agent is encrypted using the IPsec ESP tunnel mode SA described above, preventing any man-in-the-middle attack there. The need for optional confidentiality protection to mitigate the risk of compromise introduced by the protocol design was mentioned above in the threat analysis. With the link between the wireless terminal and home agent encrypted, the most likely remaining place for a successful man-in-the-middle attack is on the link between the access router and the correspondent node. Return routability has a residual vulnerability there. This vulnerability can be considerably mitigated by using the local link security mechanisms described in Chapter 5, because these mechanisms reduce the probability of a successful address hijacking or router spoofing. However, without encryption on this link, the risk cannot be completely eliminated.

Since the security provided by return routability is rather weak, the Mobile IPv6 protocol establishes rather low limits to the lifetime of route optimization bindings protected with return routability, generated keys, and nonces secured with the key on the correspondent node. Route optimization bindings secured with the generated key are limited to a lifetime of 7 minutes. Nonces are recommended to be kept active for a maximum of 3 1/2 minutes after first use, and the correspondent node is recommended to reject nonce indices for nonces over 4 minutes old. RFC 3775 also recommends that the correspondent node replace Kcn at the same time as nonces. If a Binding Update arrives with an index for a timed-out nonce, the correspondent node replies with a Binding Acknowledgement having an appropriate error code. RFC 3775 allows a fast-moving wireless terminal to reuse Ktk_H within the 3 1/2 minute window of validity so that it does not have to re-run the HoTI/HoT every time it moves, but the CoTI/CoT must always be done.

6.6   The limits of security architectures: the example of Mobile IP

One of the most difficult aspects of network security is that even if a sound security architecture is in place, it is possible to introduce security holes during the design, implementation, and even deployment phases of system development. Having a sound architecture makes it less likely that security holes will occur but security audits including threat analysis need to be undertaken during all phases of system development, including ongoing and periodic audits during deployment. These measures are needed to weed out subtle bugs, including bugs that creep in and are exploited, before their impact mushrooms.

An example of this problem can be seen in the Mobile IPv6 protocol design. The Mobile IPv6 protocol requires that certain signaling and traffic packets include routing headers with the home address or care-of address. Routing headers allow the source or destination address field on the packet to be different from the actual source or destination address to which the packet is delivered. For example, on route-optimized traffic from the wireless terminal to the correspondent node, the wireless terminal must include a Home Address Destination Option in the packet. This allows the correspondent node to match the packet to a transport layer connection sourced at the home address, even though the actual source address is the care-of address. The actual source address must be the care-of address to avoid having the packet dropped by the ingress filter on the wireless terminal's access router. These headers are a consequence of the poor integration between mobility and basic Internet routing. The headers allow this lack of integration to be overcome and for packets to be routed directly, including accomodating security measures such as ingress filtering. However, in the process, other security measures are impacted.

In the above example, when a correspondent node gets a packet with a Home Address Destination Option, the correspondent node's IP stack rewrites the packet, replacing the care-of address in the source address field with the home address from the Home Address Destination Option. The problem is that an attacker can use this process to make it look as if an attack came from a different network than the network where the attack actually originated, thereby confounding the attempts of network administrators to trace and shut off an ongoing attack. An attacker outside an administrative domain can circumvent an egress filter, by including a Home Address Destination Option in the attack traffic with an address inside the administrative domain protected by the filter, while the source address is an address outside the domain. The egress filter would typically drop an incoming packet with a source address from inside the domain it is protecting, but because the actual source address is outside the domain, the egress filter lets the packet pass. Once the victim receives the packet, it rewrites the source address with the address from the Home Address option, making it appear as if someone inside the domain has launched the attack.

To protect against these kinds of attacks, RFC 3775 places some severe limits on how the Home Address Option, the Type 2 Routing Header, and the Alternate Care-of Address Option can be used. A node processing a packet with one of these header options checks the packet against the rules, and if any are violated, the packet is dropped (and a network administrator is most likely notified in case the packet is part of an attack or misconfigured machine).

For example, the Home Address Destination Option must abide by the following rules:

- The address in the Home Address Destination Option must be a unicast routable address.
- The Home Address Destination Option must only appear once per IP header.
- The data within the Home Address Destination Option must not be altered en route to the receiver.
- If the packet is not a Binding Update, the correspondent node must only accept a packet with a Home Address Destination Option if it already has a binding for the home address in its binding cache, and the source address on the packet corresponds to the currently registered care-of address for the binding.
- If the packet is a Binding Update, the packet must be protected by a Binding Authorization Data Option. On the correspondent node, the Binding Authorization Data Option is calculated using security parameters established during return routability or by some other means. On the home agent, the Binding Authorization Data Option is protected by an ESP transport mode security association.
- The receipt of a Home Address Destination Option must not cause any change in routing or binding cache state in a node.
- If the node receiving the Home Address Destination Option does not recognize the home address and the destination address of the packet is not a multicast address, the node should return an ICMPv6 Parameter Problem message to the sender.

6.7   Summary

In this chapter, we discussed the architecture and protocols involved in IP mobility security. The standardized protocols for supporting IP mobility, namely Mobile IPv4 and Mobile IPv6, solve the problem of a host moving from one IP subnet to another, allowing the host to change its IP address in order to continue receiving packets in the new subnet. Mobile IP solves the problem by anchoring routing for the wireless terminal at a home agent that does not move, thereby allowing a correspondent host to send and receive packets tunneled through the home agent. We developed a threat analysis that detailed the threats faced by moving wireless terminals using Mobile IP. We then developed a functional architecture that aligned security interfaces with signaling and traffic interfaces involved in the mobility management protocol.

The process of mapping the Mobile IPv4 and Mobile IPv6 security protocols to the IP mobility management security architecture revealed that the Mobile IPv4 when deployed with a foreign agent does not map well to the architecture. The Mobile IPv4 security architecture with a foreign agent maps more closely to the network access control architecture from Chapter 4. The IP mobility management security architecture is a better match for deployments of Mobile IPv4 without a foreign agent and for Mobile IPv6. Consequently, subsequent discussion focused on Mobile IPv6.

Security on the interface between the home agent and wireless terminal is handled by IPsec and IKE. The design of IKE and IPsec was discussed, concentrating on IKEv2, which is a considerably simplified version of IKE, and how it is utilized for Mobile IPv6. We then described the protocol used for security on the interface between the wireless terminal and correspondent node, the return routability protocol. Return routability is a new security protocol developed specifically for Mobile IPv6 route optimization security. Return routability achieves adequate but weak authorization for route optimization binding update signaling traffic between the wireless terminal and correspondent node without requiring a global identity authentication infrastructure, which is unavailable in the Internet.

Finally, we briefly discussed the limits on the effectiveness of a security architecture in preventing problems, by examining how the design of the Mobile IPv6 protocol itself introduces the potential for a security hole. The particular problem involves the use of routing headers to rewrite the source and destination address fields on the packet. Since these address fields are used by firewalls and other filters, routing headers can facilitate propagation of attack traffic or obscure the origin of such traffic, making the job of network administrators in tracking down attacks harder. The Mobile IPv6 protocol design deals with this problem by introducing new types of routing headers and very strictly limiting their use. In general, this example points out the need for ongoing vigilance during the design, implementation, and deployment phases of system design, and even during system operation. Security bugs can crop up at any phase, and a good security architecture, while important for eliminating obvious and easily exploitable holes, is no protection against subtle bugs that can occur later in the design, implementation, and deployment process.

7 Location privacy

Users of wireless Internet services have a reasonable expectation that their activities are protected from eavesdropping and snooping by attackers even when confidentiality protection is not in use. All Internet traffic contains identifiers that allow application, transport, and network protocols to keep track of important entities and interactions. From a technical standpoint, *privacy* means that these identities are not traceable back to information allowing an eavesdropper to identify the user. If the identities are additionally masked from one or both endpoints in the protocols, then the communication is also *anonymous*. Privacy and anonymity are important security properties for certain types of transactions, and are different from confidentiality discussed in Chapter 1. The contents of a communication between two hosts can be protected by encryption to provide confidentiality from eavesdropping, while the identities of the two hosts are still exposed through unencrypted information necessary for routing. For wireless Internet communication, *location privac*y means that the geographic location of a particular wireless terminal cannot be inferred from the contents of the terminal's traffic or from unencrypted identifiers. As for general privacy, *location anonymity* means that the location is masked from endpoints as well as from eavesdroppers. Location privacy and location anonymity are issues for fixed terminals too, but because users typically carry wireless terminals with them, the risk for users is larger with wireless terminals.

In the next section, we briefly discuss the threat against general privacy of communications on the Internet and specific threats against location privacy for wireless terminals. Following that, we examine an existing, deployed approach for ensuring general privacy and anonymity of communications between two IP nodes. Then we specifically examine location privacy in systems that use IP mobility solutions such as Mobile IP and 802.11 wireless LAN. Again, we primarily focus on IPv6 since the architectural issues are clearer especially for IP mobility, but we do briefly discuss location privacy for IPv4 systems where appropriate. Finally, we show how a specific aspect of the basic Internet routing and addressing architecture, namely the forwarding algorithm used by routers, can be modified slightly to provide more location privacy. This illustrates how new architectural changes can powerfully enhance complex network systems, though the possibility of widespread deployment is often limited if the changes require extensive (and expensive) replacement of network infrastructure or terminal modifications.

7.1   Threats against privacy and location privacy

The networking stack on a wireless terminal and the applications that run on it use identifiers at various layers to map into locators. Starting at the application layer and moving down to the link layer, an example of this process is:

- The DNS name acts as an identifier for the application layer, mapping the fully qualified domain name for an IP node into the IP address to locate a correspondent node within the routing topology of the Internet.
- As discussed in Chapter 6, the IP address and port number act as a session identifier at the transport layer, mapping the session to the location of the node on which the session is running at the IP layer for routing purposes and to the actual application running on the node.
- The IP address also acts as a node identifier at the IP layer, mapping the IP address into the link layer address of the node's network interface card for last hop routing.
- Finally, in last hop routing, the link layer address acts as an identifier for a particular network interface card connected to a particular wireless terminal on the local subnet.

Since the IP address is also associated with the wireless interface card, the IP address can be viewed as identifying the interface card, and, by extension, the wireless terminal to which it is connected.

While packets are in transit between the two nodes, the identifiers are exposed to the wireless access point, routers, and, potentially, any intermediate servers like email servers or Web proxies where the traffic is queued temporarily. Even if the traffic is encrypted, intermediate entities need access to the IP header in order to route the packet, thereby exposing the identity of the two parties exchanging the traffic through correlations with the source and destination IP addresses. The exposure of identifiers to the correspondent node and intermediate nodes is basically impossible to avoid.

Exposure of identifiers becomes a threat to privacy when the identifiers can be correlated across traffic flows and used to keep track of a particular node's activities, and when the network identifiers for a node can somehow be tied back to the "real-world identity" of its user. For example, by mapping the source IP address of a node accessing a server to a user's identity, and determining what the servers do by contacting them at the IP address in the destination address, an attacker can obtain information about what the user is doing even if the traffic itself is encrypted. The user's activities involving online shopping, searching for information, accessing bank accounts, etc. could all be traced by examining the traffic and comparing the identifiers to known information about the user, a process called *traffic analysis*. A user's online life could be completely exposed to the eavesdropper. Most users have a reasonable expectation that their online activities will at least remain confidential to their terminals and possibly the servers with which they interact, and many users would prefer that the nature of their online activities also not be available to any party other than those servers.

Location information is exposed through the mapping between the IP address and the geographical deployment of IP subnets. As discussed in Chapter 6, wireless access points

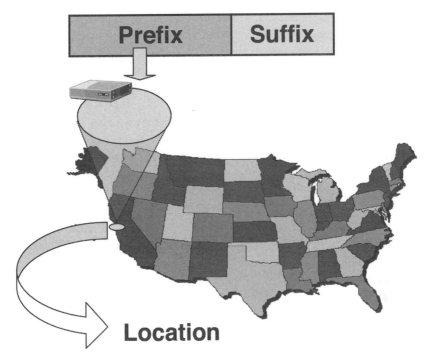

Figure 7.1 Location exposure in IPv6 addresses

are deployed to provide a particular geographical area with uniform wireless coverage from overlapping cells. IP subnets are layered on top of the wireless access points to partition IP traffic among access routers. Wireless terminals must change their IP address when they move from one subnet to another because the subnet prefix differs between the subnets. The subnet prefix thus provides an indication of a terminal's geographic location, as illustrated in Figure 7.1. The subnet prefix, which is a topological identifier, also acts as an identifier that maps to the geographic location of a particular wireless terminal.

An eavesdropper can construct a database that maps IP subnet prefixes to geographic locations. The database can be used to track a wireless terminal's movement by mapping the IP subnet prefix of the wireless terminal's current address to the geographical area where the subnet is deployed. If the attacker can correlate the IP address or some other aspect of the monitored traffic with the user's real-world identity, attacks are possible. The attacker can cyberstalk the victim as the victim moves. Depending on the motivation of the attacker, the geographical detail available, and the extent to which the attacker can locate and physically reach the victim in real time, the victim may be in direct, physical danger. Most other wireless Internet security threats do not have this aspect of immediate danger to victims, unless the wireless terminals under attack are involved in real-time control systems that manage vital infrastructure that could endanger lives on failure. The threat of location compromise is more serious than simple privacy compromise.

Unlike the systems examined in previous chapters, there are no specific interfaces between network entities with protocols requiring a location privacy architecture. Privacy, like security in general, was not a concern when the Internet architecture was first developed, so there is really no architectural support for location privacy in the Internet architecture. Because identifiers are pervasive among all protocols in the IP stack, the problem of privacy is more difficult to solve than other security problems. In order to counter the threat of privacy compromise within the existing Internet architecture, each interface where an identifier is mapped to a locator needs to be examined, and specific steps need to be taken to decouple the identifier from opportunities allowing an attacker to use the identifier to definitively locate or identify a user. We examine a few cases at the IP layer where such steps have been taken.

7.2   Security protocols for privacy in IP communication

Simple privacy requires the disruption of potential mechanisms to establish a mapping between a network identifier for an IP node and a traffic originating from the node. Usually an eavesdropper can infer something about the source of the packets if the mapping between traffic and a particular node can be established. In the worst case, the inferred information could be the actual real-world identity of the user, and the kinds of activities that they engage in on line.

7.2.1   Changing the IP Address

One way to disrupt the mapping between an identifier for an IP node and the node's traffic is to change the network identifier frequently enough that an eavesdropper has no opportunity to establish a definitive correlation. This technique can be combined with confidentiality protection to disrupt the ability of an eavesdropper to infer information about a particular user's online activities by correlating the contents of different traffic flows originating from the same node. Additional protection is possible if the network identifier can be changed in the middle of a particular transaction for the same flow. If the application uses long TCP sessions, this is typically not possible without terminating the TCP session and restarting it but for other transport protocols, or for cases where short TCP sessions are used, frequently changing the network identifier can help confuse an eavesdropper's attempts at traffic analysis. Changing the IP address does not protect against location compromise, however, location identification depends on the subnet prefix and the subnet prefix does not change if the IP node stays in the same subnet.

For example, suppose a company regularly accesses their bank account at a particular time of the month, and shortly thereafter accesses the website of an online merchant selling some products that the company uses in their business. An eavesdropper can infer from this that the firm might be transferring money in order to make a potentially large purchase. The eavesdropper can use this information for a variety of purposes. The eavesdropper can purchase a targeted DoS attack from a DoS support network (a Botnet) operator that disrupts the ability of the company to access their bank account. The DoS attack could disrupt the company's ability to do business, allowing the attacker to blackmail the company for profit. A node's IP address serves as its primary network identifier, so changing the IP address periodically – either at random or by using different IP addresses for different flows – can help disrupt the ability of eavesdroppers to perform traffic analysis.

In IPv4, this is hard to do because traditionally nodes have a single IP address per network interface card. Network operators can deploy their DHCP servers in a way that disallows a node from requesting multiple IPv4 addresses at one time. A node could release an existing address, then request a new address periodically, but many DHCP servers simply allocate the same address back to the node as was just released.

7.2.2   Privacy addresses

Changing IP addresses is easier to do in IPv6, since it is normal and expected that nodes have multiple IP addresses assigned to the same network interface card. In addition, if the network allows address autoconfiguration, the node can configure as many addresses as it needs, and change the addresses used for flows periodically, or it can occasionally configure a new address to use with a new traffic flow. RFC 4941 (RFC 4941, 2007) discusses the problem of privacy and IPv6 addresses, and recommends an algorithm for periodically changing the address. The algorithm provides the node with a randomized interface identifier, which can then be combined with the subnet prefix obtained from the Router Advertisement to form the IPv6 address. As usual, the address is then checked with duplicate address detection in the unlikely case that another node on the subnet has already claimed the address. The algorithm is particularly appropriate for interfaces that would normally use a fixed IEEE link layer address for the interface identifier, or other link layer types with a fixed link layer address. Note that if the IEEE link layer address is used as the interface identifier, an eavesdropper can not only definitively correlate traffic between flows and over time, but may also be able to trace the traffic to a particular interface card, since the IEEE link layer address is intended to be globally unique (although it can be changed) and it is assigned to the interface card on the host.

The RFC 4941 algorithm starts with a stored random value that is initialized from a pseudorandom number generator when the node boots. The steps in the algorithm are:

1. Form a 64-bit interface identifier from either the IEEE EUI-64 (64 bit) link layer identifier or the IEEE 48-bit link layer address. The techniques for forming 64-bit interface identifiers for IPv6 are described in RFC 4291, which describes the IPv6 addressing architecture.
2. Concatenate the stored random value together with the interface identifier.
3. Compute the MD5 hash of the concatenated 128-bit value.
4. Take the leftmost 64 bits of the hash and set bit six in from the left, the 'u' bit, to zero. This indicates that the interface identifier is local to the node and not universal.
5. Compare the generated identifier to identifiers that are reserved by RFC 4291 or other RFCs and to identifiers that have already been assigned on the node.

6. If the generated identifier is unacceptable, restart the process at Step 1 using the rightmost 64 bits of the hash value from Step 2.
7. Save the leftmost 64 bits as the randomized identifier.
8. Save the rightmost 64 bits as the stored random value for the next round.

If the node lacks stable storage, the random value can be generated anew for each round of the algorithm, though care needs to be taken to ensure that the random values are uncorrelated and truly random.

Cryptographically Generated Addresses (CGAs) described in Chapter 6 are also randomly generated but there are important differences that may not make CGAs appropriate for privacy purposes. Since the CGA is generated from the public key, the public key also serves as an identifier for the node. Changing the public key periodically is an approach to increasing the randomness of the address, but generating a public key is computationally expensive and the frequency may be constrained based on the capability of the node hardware and software. In any case, CGAs do not need to be certified, which helps maintain privacy, since certificates really are a way to definitively identify a node.

7.2.3   Obfuscated IP addresses

Another approach to privacy is to obfuscate the IP address so that the correspondent node and intermediate nodes cannot trace the IP address to a particular node. Having a fixed IP address to DNS name mapping is one way to ensure that your node will not enjoy privacy, since the DNS name is typically advertised globally on the Internet. Many ISPs only maintain DNS names for servers, clients typically are not provided with DNS names, though for services that require signaling from outside nodes to initiate a connection – like bidirectional, real-time voice and multimedia communication – having a DNS name may be required. In IPv4, Network Address Translation (NAT) provides weak protection, since the global IPv4 address seen by the correspondent node and network elements outside the local address realm is not the local end node address within the local addressing realm. The NAT box sitting at the border of the local addressing realm performs the mapping between the local address and global address. The protection is fairly thin, however, because most NAT boxes do not have specific privacy policy support, so they do not necessarily change the global IPv4 address periodically to avoid attempts to map a specific IPv4 address to particular flows. The node is also exposed within the local addressing realm through an unchanging local IPv4 address.

7.2.4   Onion routing

Indirect routing is another technique that helps foil traffic analysis. Onion routing (Syverson, Goldschlag, & Reed, 1998) is a system specifically designed to provide privacy by indirect routing. Figure 7.2 illustrates how onion routing works. Onion routing provides privacy and anonymity services having the following three characteristics:

- Real-time and bidirectional communication is possible, though traffic does experience increased latency due to routing indirection.

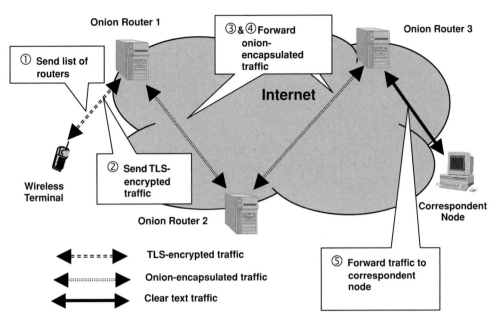

Figure 7.2 Onion routing

- The anonymous connections are independent of specific applications. Any application that uses the TCP transport protocol can use onion routing (UDP applications cannot, however).
- Onion routing does not require a centralized trust component between the nodes requiring privacy and the network components providing it. The use of secure transport connections may require certification chains and shared trust roots between different onion routers and between the onion routers and nodes.

Onion routing can be used by wireless terminals for general privacy and anonymity, and also for location privacy and location anonymity, because the correspondent node never sees the wireless terminal's IP address.

Instead of sending packets directly to the correspondent node or server, the wireless terminal initiating the connection performs the following steps:

1. The wireless terminal contacts a specialized router called an *onion router*, providing the onion router with a list of intermediate onion routers to use to deter traffic analysis.
2. The wireless router then uses a security association established with the first onion router using TLS to provide confidentiality for the traffic contents on the first hop. Since the wireless terminal's IP address is visible in the header, traffic analysis is possible on this hop, but the eavesdropper can only establish that the node is contacting an onion router, not the specific service or node on the other end of the connection.
3. The first onion router strips the contents out of the traffic. The onion router then adds headers and encrypts the traffic in multiple layers, one layer for each additional onion router through which the traffic is routed. Onion routing derives its name from

what packets look like while they are underway in the network: the multiple layers of encryption look like the layers of an onion.

4. Each onion router along the path decrypts its layer and forwards the resulting packet along to the next router.

5. When the last onion router is reached, it removes the final layer of encryption and the traffic is sent unencrypted to the correspondent node or server, with the last hop onion router's IP address as the source address.

Any attempt at traffic analysis only exposes the correlation between the last hop onion router and the server, but the identity of the originating node is protected. Onion routing also protects against compromised onion routers, because each onion router is only able to decrypt its layer. A compromised router is unable to trace the packet back beyond the previous onion router, and does not know the destination beyond the next onion router. The only network element that knows both where the packet originated and the destination is the first hop onion router.

Onion routing provides good privacy and anonymity protection; however, it does not protect against more sophisticated forms of traffic analysis such as timing analysis. Timing analysis involves using measurements of the round trip time and time between packets to derive information about a flow. Onion routing also is not useful for real-time media traffic. It only works with TCP and most media traffic uses UDP/RTP transport. Onion routing also can result in the introduction of significant end-to-end delays which are not tolerable in constant bit rate media traffic. Finally, onion routing does not protect against leakage of the IP address at higher layers. Some application protocols, such as Session Initiation Protocol which is used for setting up media sessions, can use IP addresses as endpoint identifiers.

7.3   Security protocols for location privacy in the wireless Internet

In addition to general privacy and anonymity, onion routing also provides good location privacy protection, since both the correspondent node and any eavesdroppers cannot determine the IP address of the wireless terminal that originated the traffic, so they cannot calculate a geographical mapping. If the originating terminal keeps the same set of onion routers as it changes its local IP address while it is moving, the correspondent node sees no change and therefore cannot tell that the originating terminal is moving. The only threat that onion routing will not protect against is active attacks that attempt to determine the end node's address directly, for example, an attacker that uses the *ping* program to contact the wireless terminal and learn its IP address. Any contact that does not flow through the onion router network will expose the end node's location.

7.3.1   Location privacy and Mobile IP

Mobile IP provides some location privacy protection because, like onion routing, the wireless terminal's home address does not change as the terminal changes its care-of address, as long as the wireless terminal continues to route traffic through the home agent.

RFC 4882 (RFC 4882, 2007) contains a problem statement examining the problem of location privacy for Mobile IPv6. The wireless terminal's traffic seems to originate from and terminate at the home address in the home network rather than from the care-of address. For Mobile IPv6, this means forgoing route optimization, since if route optimization is used, the correspondent node and wireless terminal engage in direct signaling when the wireless terminal changes its care-of address and the correspondent node can thereby map the home address to the care-of address. In addition, the Mobile IPv6 home agent and wireless terminal must use an IPsec ESP encrypted tunnel for all traffic in order to deny an eavesdropper the opportunity to map between the fixed home address identifying the wireless terminal and changing care-of address identifying the wireless terminal's location.

Such a correlation would allow the eavesdropper to identify the wireless terminal through the home address and track the location through the changing care-of address. For general privacy protection in IPv6, the wireless terminal must use RFC 4941 privacy addresses as described above for the care-of address so that any eavesdroppers can not establish a correlation between the IEEE or other link layer network interface card identifier used in generating the IPv6 address and the care-of address. For the same reason, privacy addresses should also be used for the home address if possible unless the home address provisioning policy of the home network operator precludes this step. If DHCP is used for IPv6 address provisioning without allowing the wireless terminal to specify the interface identifier, the wireless terminal may not control the interface identifier field in the address.

Mobile IPv4 does not support route optimization so it is not possible for the correspondent node to learn the wireless terminal's location through the care-of address. But Mobile IPv4 also does not support an encrypted tunnel between the foreign agent or the wireless terminal (if the care-of address is co-located on the wireless terminal), so it is possible for an eavesdropper in the wireless terminal's access network to establish a correlation between the home address and care-of address and thereby track the identity and location of the wireless terminal. Since Mobile IPv4 has mostly been deployed in cellular networks, this vulnerability has not appeared as an issue because cellular network operators maintain tight control over their access networks making compromise from outside agents unlikely (though compromise from insiders is always a possibility). On less controlled networks, such as 802.11 hotspots, eavesdropping is a more potent threat.

7.3.2   Problems with home agent tunneling for location privacy in Mobile IP

A residual, minor vulnerability exists within the home network itself. When the wireless terminal is away from the home network, the Mobile IPv6 home agent responds to Neighbor Solicitations to resolve the wireless terminal's IPv6 address to a link layer address. This process is called proxy Neighbor Discovery, because the home agent is acting as a proxy for the wireless terminal. Any terminal on the home network that receives a proxy Neighbor Advertisement from the home agent will immediately be able to identify that the wireless terminal is no longer on the home network. The terminal can compare the returned link layer address with the link layer address for the home agent, and, if the two match, the terminal knows that the wireless terminal is not on the home network. If the wireless terminal treats nodes on the home network exactly like any other correspondent node, however, and continues to route traffic through the home agent, the care-of address is not exposed and the home network terminals are unable to determine where the wireless terminal is located. Mobile IPv4 has the same vulnerability, through proxy ARP.

As with onion routing, routing all traffic through the home agent can result in significant increases in end-to-end delay, which may negatively impact the performance of real-time media traffic. Note also that this solution only covers exposure of identity and location at the IP layer. If an application layer identifier such as a DNS name is bound to the home address, the correspondent node may be able to deduce the wireless terminal's identity through the DNS name. The correspondent is unable to track the wireless terminal's location, however.

Other identifiers that can be somehow tied to the home address and remain constant over longer time periods also expose the wireless terminal's location to compromise. For example, the home agent and correspondent node use an IPsec identifier, the Security Parameters Index (SPI), on the ESP packets to identify the security association. The SPI is another point of exposure. After the IPsec security association is established between the home agent and wireless terminal, the SPI typically does not change until the SPI times out (and even then it may be renewed). The SPI therefore provides a fixed identifier by which an eavesdropper in the access network could track the movements of the wireless terminal. Unlike the home address, however, the SPI typically has no long-term correlation with an identity and an eavesdropper would require additional information to make the connection between the SPI and some other information exposing the wireless terminal's identity. Such information could be obtained by monitoring the initial IKE exchange between the home agent and wireless terminal, but since the number of packets exchanged is very limited, the window of exposure for the wireless terminal is very narrow. The eavesdropper must be correctly positioned at just the right place and time when the wireless terminal boots up and establishes the IPsec security association using IKE in order to exploit this vulnerability. An eavesdropping attack could additionally be foiled by periodically changing the IPsec security association with the home agent, thereby causing the SPI to change. Establishing a security association is typically a time-consuming process, so this option may not be available to a wireless terminal.

Finally, like onion routing, Mobile IP provides no protection against an active attacker. An attacker that is scanning an access network using *ping* can obtain the wireless terminal's care-of address. This may allow the attacker to map the care-of address back to the home address, thereby establishing the wireless terminal's location.

7.3.3   Location privacy and access network link layer identifiers

Indirect routing provides location privacy at the IP layer, but the location of a wireless terminal can also be exposed at the link layer by the link layer address. A wireless terminal typically does not change its link layer address when it moves from one access point to another, unlike the care-of address at the IP layer. In many cases, the link layer address definitively identifies the network interface card in the host. For example, in IEEE 802.11 networks, the 802.3 link layer address is typically programmed into the network interface hardware at the factory (Wikipedia, 2008c). It contains fields identifying the manufacturer and globally identifying the specific interface card. While it is possible to dynamically change the link layer address, the address is typically not changed, and, in fact, network access authentication using 802.1x/802.11–2007 (802.11, 2007) requires that the link layer address not change since the security association between the access point and wireless terminal is identified by the link layer address. The wireless terminal can change its link layer address when it moves between access points, but this may require a lengthy reauthentication procedure. From the network's standpoint, a wireless terminal with a different link layer address looks like a newly arrived entrant.

Cellular networks handle this problem by using the wireless terminal's global identifier for network access authentication, then assigning a temporary identifier for further use to hide the host's identity. For example, in GSM networks, the wireless terminal uses its International Mobile Subscriber Identity (IMSI) on first access (Wikipedia, 2008d). The network assigns the terminal a Temporary International Mobile Subscriber Identity (TIMSI) after the terminal has successfully navigated network access authentication, which is used for further communication with the network. The window of vulnerability for exposing the wireless terminal's identity is very limited, reducing the risk of identity compromise.

Network operators have two deployment and operational tools that they can use to reduce the vulnerability of their clients' wireless terminals to location privacy compromise:

- By increasing the geographical coverage area of access routers and subnets, the access network subnet prefix provides coarser-grained information about where the wireless terminal is located. For example, the care-of address in an access network that assigns a subnet to each floor in an office building provides an attacker with much more detailed information about the location of a wireless terminal than a care-of address in an access network with a subnet the size of a large metropolitan area.
- By frequently changing the subnet prefix to geographical mapping in access network through network renumbering, any mapping between subnet prefixes and geographical areas is limited in duration. An attacker that establishes such a mapping will find that its usefulness is time limited.

Access network renumbering is difficult in IPv4, and probably not a practical tool for location privacy, since there are no protocols for automating it. IPv6 has some support for network renumbering, but even in IPv6, renumbering is likely to be challenging. Since network renumbering invalidates all the addresses in the access network, any wireless terminals that are running IP stacks must shut down IP service and re-establish a new IP address. Therefore, network renumbering is probably only feasible on a long-term basis.

7.3.4   Protocol solutions to location privacy in Mobile IPv6

The above solutions to location privacy involve utilizing Mobile IPv6 or deploying networks in particular ways that deny the correspondent node and an eavesdropper the ability to establish a correlation between the care-of address and the home address. Without such a correlation, traffic analysis cannot be used at the network layer to track the wireless terminal's location. The price for location privacy, however, is the lack of ability to do route optimization. This price may be too steep for some applications, for example real-time voice and multimedia, or for cases where the correspondent node and wireless terminal are in the same location, but the wireless terminal's home agent is on the other side of the world. The latency due to indirect routing through the home agent may be too long for good performance.

The route optimization signaling between the wireless terminal and correspondent node exposes the correlation between the home address and care-of address in two ways: when the binding moves from the home address to the care-of address and when it moves back again. In most cases, the route optimization signaling is sent in clear text, so the correlation between the home address and the care-of address is clearly visible to an eavesdropper. A proposal has been made for modifications to the Mobile IPv6 protocol to allow route optimization but to remove the ability of an eavesdropper to establish a correlation between the care-of address and home address (Qiu, Zhao, and Koodli, 2007). The proposal contains the following approaches:

- Substitute a nonroutable identifier for the home address as the endpoint identifier in route optimization traffic between the correspondent node and wireless terminal. The nonroutable identifier is generated by encrypting the home address using a shared key established through an extension of the return routability protocol. This approach hinges on the observation in Chapter 6 that the home address really only serves as an endpoint identifier for route optimized traffic, the care-of address is the routing locator.
- Replace the home address with a routable address generated between the home agent and the wireless terminal, but with an interface identifier that is generated in various ways to reduce the ease of discovering the wireless terminal's identity. Replace the home address periodically with an RFC 4941 privacy address.

Note that this proposal is still under study by the IETF, and has not yet been standardized.

By replacing the home address, route optimization is possible without exposing the care-of address and home address in one packet, so an eavesdropper cannot establish the correlation. While all three approaches disrupt the ability of eavesdroppers to establish a correlation between the home address and care-of address, the correlation is impossible to deny the correspondent node if the replacement home address is nonroutable, since the correspondent node must participate in the protocol that establishes the nonroutable identifier. Routable home address replacements obscure the actual home address, but only if the routable replacement was used to originally establish the traffic flow between the wireless terminal and correspondent node. The protection provided by a routable home address is also limited, since an eavesdropper can deduce the home network from the subnet prefix and can possibly map that to other constant identifiers in the wireless terminal's traffic, like the IPsec SPI or application layer identifiers. Also, none of these approaches protect against active attacks, in which the attacker pings the wireless terminal periodically to obtain the care-of address, and uses that to track location.

7.4   An architectural approach to location privacy

Taking a step backward and approaching the problem of location privacy from an architectural standpoint, there are two architectural factors facilitating location compromise in wireless IP networks:

- The correlation between the subnet prefix and the geographic location of the access points and routers that provide wireless connectivity to the subnet in a wireless network deployment.
- The existence of fixed identifiers in the traffic stream that allow an eavesdropper to deduce a wireless terminal's identity and location by matching an identifier to the subnet prefix from a particular location.

The solutions described above address the second factor on a case by case basis by decoupling the ability of an eavesdropper to establish a correlation between the host's identity and location. These solutions do not make any fundamental architectural changes, they just change the way networks are deployed, or how a wireless terminal uses existing protocols, to reduce the opportunities available to eavesdroppers for location compromise. In a few cases, changes in existing protocols have been proposed to hide the identity of the wireless terminal, but these changes do not stem from any basic architectural changes.

A solution addressing the first factor seems more difficult to achieve, since the correlation between the subnet and the geographic location is a function of the constraints imposed by the underlying technologies of wireless access and IP. Access points localize wireless Internet access geographically and the IP routing and addressing architecture localizes hosts topologically. Access points are connected into stub subnets identifiable by the forwarding algorithm through the subnet prefix. The routing and addressing architecture requires that the subnet identifier be present on any end-to-end routed packet so that the forwarding algorithm can route packets to their proper destination and identify the source of the packets. Any solution decoupling the subnet identifier from geographical location requires either a fundamental change in the underlying wireless access technology or in the IP routing and addressing architecture.

In the next sections, we examine an experimental architectural change in IPv6, called Cryptographically Protected Prefixes (CPP), that decouples the subnet identifier from geographic location by masking the actual subnet identifier (Trostle *et al.*, 2005). This solution modifies the basic routing and addressing architecture by changing the forwarding algorithm used in IP routing. While such architectural modifications can provide considerably more powerful and consistent solutions than patching the ability to do identifier mapping on a case by case basis, the likelihood that they will be widely deployed is unfortunately minimal. Deep architectural changes typically require deep changes in or even replacement of network equipment that most wireless access providers already have paid to deploy. Unless the incentive for deep changes or penalty for not making such changes are sufficient, network providers are unwilling to invest in the equipment and deployment cost involved in changing the basic infrastructure. Nevertheless, while the cases examined in previous chapters have applied architectural analysis to existing IP protocols and systems, CPP illustrates how new architectural changes can be introduced in a systematic way.

7.4.1　The Cryptographically Protected Prefix (CPP) algorithm

CPP is motivated by the observation that the ability to map a geographic location to an IP address in IPv6 originates with the need for the IPv6 access routers to advertise fixed subnet prefixes identifying the subnet(s) available through the access points serving the geographic location. These prefixes are then used by IPv6 nodes on the subnet to autoconfigure addresses. If DHCP is used, the router may not advertise the prefixes, but the DHCP server will still maintain a collection of fixed prefixes for addresses that it provisions to nodes on the subnet.

If these prefixes remain constant over long periods of time, an attacker can collect information on the subnet prefix to geographical location mapping, and use that information to identify the location of wireless terminals sending IP packets from the subnets. The solution that CPP applies to this problem is to mask the subnet prefix so that there is no fixed subnet prefix for a particular geographic location. Instead, each wireless terminal has an independent subnet prefix, uncorrelated with the prefixes of other hosts on the same subnet. If the wireless terminal also uses a randomly generated interface identifier, by changing the interface identifier periodically through generating RFC 4941 privacy addresses or through periodically generating CGAs from different RSA keys, the address cannot be traced to a geographic area and cannot be mapped to a particular wireless terminal.

CPP operates by encrypting part of the IPv6 subnet prefix used for intradomain routing. As discussed in Chapter 5, the IPv6 address typically consists of a 64-bit subnet prefix and a 64-bit subnet identifier. As shown in Figure 7.3, however, the IPv6 address can be broken down into three parts:

- The first portion of the subnet prefix, typically 16 bits long, that is used for interdomain routing on the Internet. This portion is designated P_0 in Figure 7.3.
- The rest of the subnet prefix, typically 48 bits, used for intradomain routing among subnets and subdomains serviced within the primary routing domain. This portion is designated P_S in Figure 7.3.
- The 64-bit interface identifier, which identifies a particular network interface card on a host. The interface identifier is designated M_i in Figure 7.3.

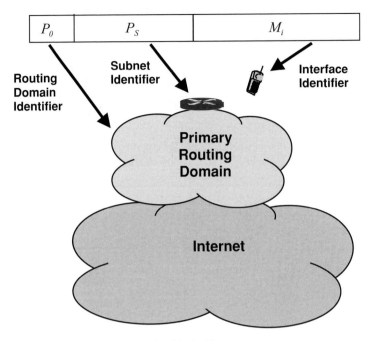

Figure 7.3 Detailed structure of an IPv6 address

CPP identifies the 16-bit P_0 portion with an *address privacy domain*, in which the following 48 bits in P_S do not directly identify the subnet. Instead, the actual subnet is determined by decrypting portions of the address successively as packets are routed through the primary routing domain. The decryption operation requires changes in the forwarding algorithm on routers in the address privacy domain. The interdomain routing prefix cannot be masked because it must be used by border routers connected to the Internet in order to find the right primary routing domain, and many of those routers will not support the privacy domain enhancement.

CPP depends on a feature of IPv6 route advertisement called *route aggregation*. In a routing domain that utilizes route aggregation, routers inside the routing domain advertise summarized subnet prefixes to routers topologically closer to the border router. The prefixes are summarized by dropping the rightmost bits until the remaining prefix matches the summarized prefixes of all other routers at that topological level. Figure 7.4 shows an example of route aggregation. In the figure, P_0 is the interdomain prefix, while the P_{Si} are bit fields corresponding to portions of the intradomain subnet prefix that are summarized away as the route advertisements progress up the topology to the border router. At the bottom of the hierarchy, the access routers advertise subnet prefixes to hosts on the stub subnets. So, for example, the access router on the left side advertises $P_0 P_{S1} P_{S2} P_{S4}$ to hosts on the stub subnet but summarizes the route, together with the next access router to the left which is on the same subnet, into $P_0 P_{S1} P_{S2}$, and so on up to the border router, which advertises the interdomain prefix, P_0, on the Internet.

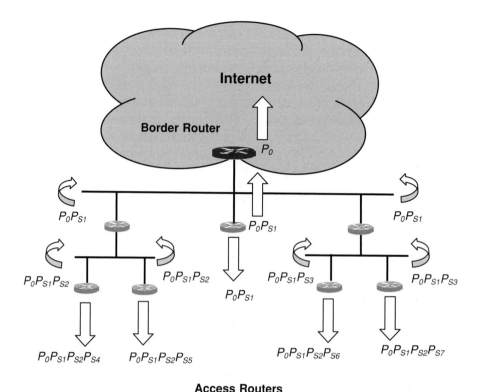

Figure 7.4 An example of route aggregation

Route aggregation is primarily a tool for reducing the size of the routing tables within a routing domain, and, more importantly, on border routers connected to the Internet. So, for example, if route aggregation is not used in Figure 7.4, the routers above the access level have to maintain all the subnet prefixes advertised by all the access routers in their routing tables. Complete route aggregation, as shown in Figure 7.4, is necessary for the simple version of CPP presented here to work properly. Complete route aggregation is not always possible, since it requires a hierarchically organized routing topology. More complex versions of CPP allow less aggregation (Trostle *et al.*, 2005).

The standard IP forwarding algorithm uses longest prefix matching to determine what the next hop towards the destination should be. Figure 7.5 illustrates the standard IP forwarding algorithm. Packets incoming from the Internet are forwarded to their end destination by matching the subnet prefix against successively more finely detailed advertised prefixes in the routing tables of routers along the path. A router on the path matches the destination address subnet prefix from an incoming packet against all the prefixes in its routing table. The prefix that matches the most bits in the address indicates which router should be the next hop. The router extracts the next hop information from the routing table, and forwards the packet through the interface indicated in the next hop information, to the next router on the path.

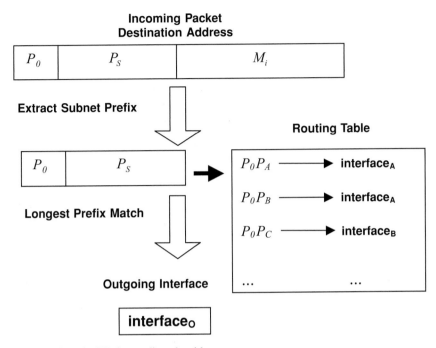

Figure 7.5 Standard IP forwarding algorithm

CPP modifies the standard IP forwarding algorithm by adding a step prior to longest prefix matching. Figure 7.6 illustrates the CPP forwarding algorithm. All routers advertising a specific summarized prefix upward share a *level key*. The route lookup procedure combines the level key and the interface identifier of the address to unmask the particular bit field in the subnet prefix of the packet's destination address that identifies the route to the next hop on the next level. The masked subnet prefix bits, X_S, are uncorrelated for wireless terminals on the same subnet, ensuring location privacy. For packets incoming from the level above, the router uses the level key and interface identifier to unmask the unique bits identifying the next hop at the next aggregation level. After the bit field has been unmasked, the longest prefix matching algorithm is applied using the prefix to look up the next hop, exactly as in the basic IP forwarding algorithm. The bits that are still masked do not influence the matching, because they are not relevant to longest prefix matching even in a clear text (completely unmasked) prefix. Forwarding of the packet to the next hop proceeds like the basic forwarding algorithm. But since each router can only unmask the bits identifying the route to the next level, the wireless terminal's location privacy is protected from compromise of routers or eavesdropping up to the hop immediately preceding the access router. It is only at this hop that the full subnet prefix becomes unmasked.

Since routers can only unmask bits identifying the next hop, they require access to a partially unmasked address in order to have access to the bits in the prefix for higher aggregation levels. The subnet prefix with unmasked fields is encrypted with a key shared between the forwarding router and the next hop router, and included in an IPv6

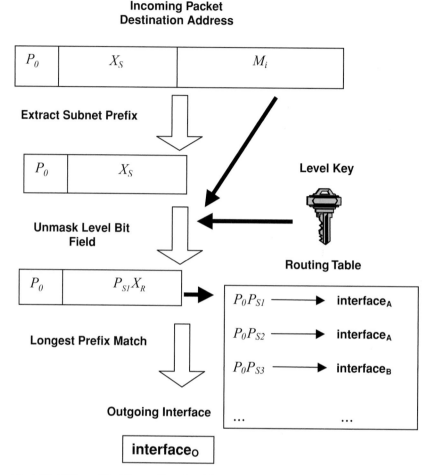

Figure 7.6 CPP-modified forwarding algorithm

header option in the outgoing packet. This allows a next hop router on a lower level to obtain the clear text address up to the level bit field of the previous hops. If the unmasked prefix is not included, the next hop router is only able to unmask the particular bit fields that correspond to the next level, and the longest prefix matching step fails. Encrypting the partially unmasked address in transit decreases the risk that an eavesdropper can obtain information about the wireless terminal's location from the clear text bits in the address.

The above algorithm describes how packets incoming from the Internet are routed to hosts within the privacy domain. Packets routed between hosts in the location privacy domain require another modification to standard IP forwarding. Normally, standard IP forwarding only sends packets destined for the Internet on a default route out of the local routing domain via a border router. This procedure is also used with CPP. In standard IP routing, however, packets destined for hosts within the local routing domain are routed directly using the standard IP forwarding algorithm. In the standard IP forwarding algorithm, a longest prefix match at an intermediate router within the domain registers a hit for another router within the routing domain, and the packet is forwarded there. With CPP, the standard IP forwarding algorithm cannot be used for destinations within the address privacy domain, because intermediate routers do not have the ability to unmask the entire subnet prefix. As a result, packets destined within the address privacy domain are also forwarded from the access router to a border router. The border router recognizes from P_0 that the destination is within the local routing domain and forwards the packet back down the route aggregation hierarchy allowing the address to be decoded. The resulting routes are longer than in standard IP routing, but most routing domains do not have many hops to the border router to ensure that traffic to and from the Internet does not experience substantial routing delays, so the additional routing delay should not be excessive. In addition, some refinements to CPP allow shortcuts in internal routing (Trostle *et al.*, 2005).

An additional factor in location privacy is IPv6 Neighbor Discovery and address autoconfiguration. Since Neighbor Discovery and address autoconfiguration only work on the local subnet, the receipt of a Neighbor Discovery or address autoconfiguration message by a terminal compromises the address privacy of the sender, because it informs the receiver that the sender is on the same subnet, and therefore in the same geographic area, as the sender. CPP therefore requires that terminals do not perform IPv6 Neighbor Discovery or address autoconfiguration. Instead, terminals request addresses from the access router. This allows the access routers to additionally control the address allocation so that terminals receive properly masked addresses. The access routers also record the link layer address of terminals requesting IPv6 addresses, so that the access routers do not need to do Neighbor Solicitation when a packet arrives for the terminal. This avoids the need for the aspects of Neighbor Discovery that expose a terminal's location on the subnet and are problematic for location privacy. Terminals still use IPv6 router discovery to find access routers, but access routers do not insert subnet prefixes into Router Advertisements, since clear text subnet prefixes would leak location information. In addition, the Router Advertisements must indicate that stateful address configuration is used on the subnet.

7.4.2   Key and address provisioning for CPP

CPP requires two sets of keys provisioned on routers within the location privacy domain:

- A pairwise key shared between next hop routers, which is used to encrypt the partially decrypted address.
- The level keys shared between all routers that advertise the same summarized prefix, which is used to decrypt the bits in the subnet prefix corresponding to the next level down in the route aggregation hierarchy.

The provisioning of keys and addresses having masked subnet prefixes is controlled by a key distribution and masked prefix address server. This server has a security association with every router in the location privacy domain allowing the server to provision sensitive keys and masked prefix addresses to the routers with confidentiality and authenticity ensured. The server must provide the masked prefix addresses rather than letting the routers calculate them because the server is the only network entity with access to all the level keys. The server sends blocks of topologically correct addresses with masked prefixes to the access routers, and the access routers then provision the addresses on request to the wireless terminals. Because the number of bits at each level is very limited, the possibility of guessing attacks on a particular level is relatively high, so the key server must push out new level keys and address blocks periodically, and wireless terminals must renew their address leases periodically to obtain a new masked prefix address. This keeps the keys fresh and the addresses new so the time window of exposure to guessing attacks is limited.

7.4.3   Residual vulnerabilities in CPP

There are two points of residual location privacy vulnerability in CPP:

- The key distribution and masked prefix address server is the most vulnerable entity because it has access to all the level keys and masked prefix addresses.
- The access routers and their immediately adjacent parents upward in the route aggregation hierarchy need access to clear text routing prefixes in order to perform longest prefix matching.

The compromise of a router above the level of an access router's parent also exposes some location information, but the granularity of the exposed information is relatively high, so the wireless terminals' location information still achieves some level of protection.

These residual vulnerabilities are difficult to eliminate. The key distribution and masked prefix address server can be deployed in a way that is "radiation hardened," i.e. by taking deployment and administrative steps that significantly reduce the probability of compromise or downtime due to DDoS attacks. For example, access to the key distribution server can be strictly controlled, including not using a globally routable IP address for the server, and the server can be replicated so that a replica can be brought up quickly if the primary server is subject to a DDoS attack. But the key distribution and masked prefix address server remains a target for attackers, and therefore must be administered and monitored carefully.

The access routers and their immediately adjacent neighbors are harder to control, however. Because CPP requires that the matching next hop prefix in the router table is in clear text exactly as for the basic longest prefix matching algorithm, routing information is still exchanged between routers in clear text (encrypted in transit of course to ensure confidentiality from eavesdroppers), so compromise of an access router exposes all wireless terminals on the subnet, and compromise of a router one level above an access router exposes all terminals on all access routers below that level. Again, deployment measures can help mitigate the risk. A few possible measures are strictly controlling access to the access routers and upper level intermediate routers and having the access router parents cover only a few access routers to limit the amount of damage should an access router's parent be compromised.

7.4.4 Functional architecture for CPP

Given the above description of the CPP algorithm, the next step is to develop a functional architecture. Figure 7.7 contains a summary architecture diagram illustrating the functions, functional entities, and the interfaces between them. From the algorithm described above, there are four functional entities in CPP:

- the Key Distribution and Masked Prefix Address Server
- the Location Privacy Domain Router
- the Access Router
- the Mobile Node.

A functional architecture supports the following interfaces between these entities:

- CPP1: the interface between the Key and Masked Prefix Address Server and the Location Privacy Domain Routers.
- CPP2: the interface between Location Privacy Domain Routers themselves.
- CPP3: the interface between the Access Router and Mobile Node.

Since the Mobile Nodes only communicate with the Access Router, there is no need to define an interface between the Mobile Node and the key distribution and masked prefix address server, for example. Also, note that this characterization applies to functional entities and the interfaces between them only, not the actual network entities (i.e. implemented boxes with software). For example, an Access Router network entity must support all three interfaces, while other routers in the location privacy domain only need to support CPP1 and CPP2.

There are three functions that are not associated with network interfaces:

- the Masked Address Block Generation function in the Key Distribution and Masked Address Server;
- the Level Key Generation and Management function in the Key Distribution and Masked Address Server;
- the Unmask Address function in the Location Privacy Domain Router.

These functions are part of the programmatic interfaces internal to the respective network entity implementations. They are included here because they are essential to the design of CPP, but in an interoperability specification for a network protocol, they would typically be defined only in abstract terms, with enough detail to ensure that the network protocols on the various interfaces could be adequately specified. The actual details of the implementation are specific to the particular programming platforms on which CPP is implemented.

Note that Figure 7.7 does not include the functions and interfaces necessary to establish security associations between the Location Privacy Domain Routers, between the Key and Masked Address Server and a Location Privacy Domain Router, and between the Access Router and a Mobile Node. The exact nature of these interfaces and the functions depend on the cryptographic and key distribution algorithms used to implement the security associations. Additional functional entities, like a certificate authority or AAA

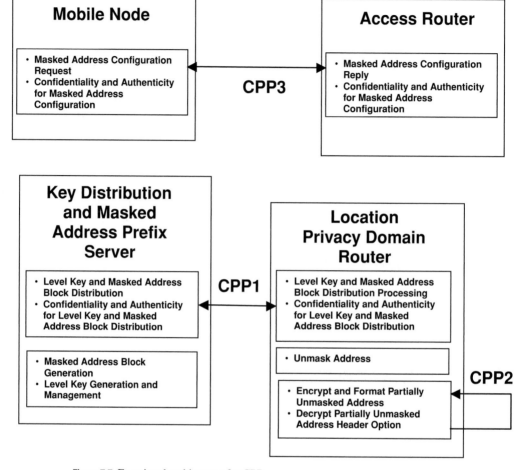

Figure 7.7 Functional architecture for CPP

server, could be introduced depending on the cryptographic algorithm. While these details are important for an actual system design, they are not necessary for understanding how to develop a functional architecture for a new architectural modification such as CPP.

In the following sections, we define the functions associated with the functional entities in the figure.

7.4.5   CPP Key Distribution and Masked Address Prefix Server functions

Table 7.1 contains a list of the Key Distribution and Masked Address Server functions, the security services they implement, the parameters for the functions, and the objects returned by the functions.

Table 7.1 Key Distribution and Masked Address Prefix Server functions

Function	Security services	Parameters	Return
Level Key and Masked Address Block Distribution	– Key and masked prefix address provisioning to Location Privacy Domain Routers – Management of secure signaling with Location Privacy Domain Routers	– IP address for a specific Location Privacy Domain Router – Subnet information for the Location Privacy Domain Router (empty if no address block required) – Level keys for the Location Privacy Domain Routers, identified by router IP addresses – Security association, including keys and other parameters, between the Key and Masked Address Server and the specific Location Privacy Domain Router	– None on success, error indication if an error occurred during processing
Masked Address Block Generation	– Masked prefix address generation	– Level keys for the location privacy domain – Subnet information for the Location Privacy Domain Router – Number of required addresses in address block	– A block of addresses with masked subnet prefix for distribution to the Location Privacy Domain Router
Level Key Generation and Management	– Key generation and management	– Route aggregation and topology information for the location privacy domain	– Level key table for the location privacy domain
Confidentiality and Authenticity for Level Key and Masked Address Block Distribution	– Data origin authentication and confidentiality protection on key and masked prefix address provisioning messages	– Security association, including keys and other parameters, between the key and masked address server and the specific Location Privacy Domain Router. – Clear text Level Key and Masked Address Block Distribution Message on send – Encrypted, authenticated Level Key and Masked Address Block Reply on receive	– Encrypted and authenticated Level Key and Masked Address Block Distribution Message on send – Clear text, verified Level Key and Masked Address Block Reply from Location Privacy Domain Router on receive or an indication of security failure if the message did not decrypt or verify

Level Key and Masked Address Block Distribution Message Formulation function

The Level Key and Masked Address Block Distribution Message Formulation function formulates and sends a Level Key and Masked Address Block Distribution message to a particular Location Privacy Domain Router. If the router is an Access Router, the subnet prefixes on the stub subnet where hosts attach are provided, indicating that an address block for the router should also be generated, and sent along with the level key. If the router is not an Access Router, no subnet information is provided, and just a level key is sent. The function has no return value if successful; otherwise an error indication is returned identifying the source of the error.

Masked Address Block Generation function

The Masked Address Block Generation function generates masked addresses for a particular Access Router. It takes as parameters the level keys for the location privacy domain, the subnet information on the stub subnet serviced by the Access Router where hosts attach, and the number of desired addresses in a block. The function returns the masked prefix address block.

Level Key Generation and Management function

The Level Key Generation and Management function generates level keys if none exist or retrieves an existing level key table. It takes as parameters the route aggregation and topology information on the location privacy domain and returns a table of level keys for the location privacy domain.

Confidentiality and Authenticity for Level Key and Masked Address Block Distribution function

The Confidentiality and Authenticity for Level Key and Masked Address Block Distribution function maintains confidentiality and authenticity on transmission of the level key and masked address block distribution to a particular Location Privacy Domain Router. On send, it takes a Level Key and Masked Address Block Distribution message in clear text and returns an encrypted and authenticated message for sending to the Location Privacy Domain Router. On receive, it takes an encrypted, authenticated message from the location privacy router acknowledging receipt and returns a clear text, verified message. If an error occurs on processing the acknowledgement, the function returns an error indication.

7.4.6   CPP Location Privacy Domain Router functions

Table 7.2 contains a list of the functions supported by the Location Privacy Domain Router, the security services they implement, the parameters for the functions, and the objects returned by the functions. The table includes functions involved in both intra-location privacy domain routing on CPP2 and interaction with the key and masked address prefix server on CPP1.

Level Key and Masked Address Block Distribution Processing function

The Level Key and Masked Address Block Distribution Processing function processes a provisioning message from the Level Key and Masked Address Block Server to obtain the level key for this router's level and, if the router is an access router, a block of

Table 7.2 Location Privacy Domain Router Functions

Function	Security services	Parameters	Return
Level Key and Masked Address Block Distribution Processing	– Key and masked prefix address provisioning from Key and Masked Prefix Address Server – Management of secure signaling with Key and Masked Prefix Address Server	– Encrypted and authenticated Level Key and Masked Address Block Distribution message – Security association, including keys and other parameters, between the Key and Masked Address Server and the Location Privacy Domain Router	– Level key for this aggregated routing level – Masked Address Block if access router, otherwise empty
Unmask Address	– Decryption of masked address prefix	– Level key for this routing aggregation level – Partially unmasked destination address	– Destination address with next-hop bits unmasked
Confidentiality and Authenticity for Level Key and Masked Address Block Distribution	– Data origin authentication and confidentiality protection on key and masked prefix provisioning messages	– Security association, including keys and other parameters, between the key and masked address server and the Location Privacy Domain Router – Clear text Level Key and Masked Address Block Distribution Message reply on send – Encrypted, authenticated Level Key and Masked Address Block Distribution Message on receive	– Clear text, verified Level Key and Masked Address Block Distribution Message on receive or an indication of security failure if the message did not decrypt or verify – Encrypted and authenticated Key and Masked Address Block Reply send
Encrypt and Format Partially Unmasked Address	– Encryption of partially unmasked address for the next hop router at the next aggregation level	– Partially unmasked destination address – Security association, including keys and other parameters, between this router and the next hop router	– Encrypted and authenticated IPv6 header option containing the partially unmasked address
Decrypt Partially Unmasked Address Header Option	– Decryption of partially unmasked address from the previous hop router at the previous aggregation level	– Encrypted and authenticated IPv6 header option containing the partially unmasked destination address – Security association, including keys and other parameters, between this router and the previous hop router	– Clear text, verified IPv6 header option with partially unmasked destination address

addresses for provisioning Mobile Nodes. The input parameters are an encrypted and authenticated Level Key and a Masked Address Block Distribution message from the server and the security association with the server. The function returns the level key and, if provided in the input, a block of masked prefix addresses.

Unmask Address function

The Unmask Address function takes a destination address partially unmasked up to the current level and returns a destination address with the prefix bits unmasked for routing at this level.

Confidentiality and Authenticity for Level Key and Masked Address Block Distribution function

The Confidentiality and Authenticity for Level Key and Masked Address Block Distribution function maintains confidentiality and authenticity on transmission of the level key and masked address block distribution to a particular Location Privacy Domain Router. On send, it takes a Level Key and Masked Address Block Distribution message in clear text and returns an encrypted and authenticated message for sending to the Location Privacy Domain Router. On receive, it takes an encrypted, authenticated message from the location privacy router acknowledging receipt and returns a clear text, verified message. If an error occurs on processing the acknowledgement, the function returns an error indication.

7.4.7   CPP Mobile Node functions

Table 7.3 contains a list of the Mobile Node functions, the security services they implement, the parameters for the functions, and the objects returned by the functions. In addition to the functions that are new or modified from a standard IPv6 host, CPP requires that a compliant Mobile Node drop certain functions that are part of the base IPv6 host specification, in particular Neighbor Discovery Solicitation/ Advertisement.

Note that the Confidentiality and Authenticity for Masked Address Configuration function may also require a preexisting security association with the Access Router, if a shared key algorithm is used. If a public key algorithm is used, a preexisting security association, in the form of a certificate exchange, may not be required if the initial trust relationship between the network and the Mobile Node is established at the link layer.

Masked Address Configuration Request function

The Masked Address Configuration Request function requests the provisioning of a masked prefix IPv6 address from the Access Router whose link local IPv6 address is an input parameter, and returns the result, a globally routable IPv6 address with masked subnet prefix, to the IP stack address provisioning module. If an error occurred during the processing of the request or reply, the function returns an error indication.

Confidentiality and Authenticity for Masked Address Configuration function

The Confidentiality and Authenticity for Masked Address Configuration function establishes confidentiality and authenticity on the masked address traffic between the Access Router and the Mobile Node. On send, the function takes the security association between

Table 7.3 CPP Mobile Node functions

Function	Security services	Parameters	Return
Masked Address Configuration Request	– Masked address configuration signaling	– Link local IPv6 address of access router	– Masked IPv6 address usable for configuring an IPv6 network interface or error indication
Confidentiality and Authenticity for Masked Address Configuration	– Data origin authentication and confidentiality protection on address configuration signaling	– Security association, including keys and other parameters, between the Access Router and the Mobile Node – Clear text Masked Address Configuration Request on send – Encrypted and authenticated Masked Address Configuration Reply from router on receive	– Encrypted and authenticated Masked Address Configuration Request on send – Clear text, verified Masked Address Configuration Reply from Access Router on receive or an indication of security failure if the message did not decrypt or verify

the Access Router and Mobile Node and the clear text Masked Address Configuration Request as parameters and returns an encrypted and authenticated request message. On receive, the function takes the security association between the Access Router and Mobile Node and the encrypted and authenticated Masked Address Configuration Reply from the Access Router and returns a clear text, verified response appropriate for use in address configuration on the Mobile Node. If the response does not decrypt or verify, an error indication is returned.

7.4.8   CPP Access Router functions

Table 7.4 contains a list of the Access Router functions, the security services they implement, the parameters for the functions, and the objects returned by the functions. Note that these functions only apply to the CPP3 interface. The Access Router additionally supports functions in the CPP1 and CPP2 interfaces. The sections below contain more detail on the functions. CPP requires that a compliant Access Router also drop Neighbor Discovery Solicitation/Advertisement as for the Mobile Node. In addition, the Access Router must continue to support router discovery, but must advertise stateful address configuration and must not insert any subnet prefixes into the Router Advertisements. Since each Mobile Node receives a separate, customized subnet prefix, subnet prefix advertisements in the Router Advertisements are not necessary, and if they were provided in clear text, they would leak location information.

Table 7.4 CPP Access Router functions

Function	Security services	Parameters	Return
Masked Address Configuration Reply	– Masked address configuration signaling	– Encrypted, authenticated Masked Address Configuration Request message from the Mobile Node	– No return on success, error return if any error in processing request or reply occurs
Confidentiality and Authenticity for Masked Address Configuration	– Data origin authentication and confidentiality protection on address configuration signaling	– Security association, including keys and other parameters, between the Access Router and Mobile Node – Encrypted Masked Address Configuration Request from Mobile Node for receive – Clear text Masked Address Configuration Reply to Mobile Node for send	– Clear text, verified Masked Address Configuration Request from Mobile Node for receive or an indication of security failure if the message did not decrypt or verify – Encrypted and authenticated Masked Address Configuration Reply to Mobile Node for send

Masked Address Configuration reply

The Masked Address Configuration Reply function replies to a request from a Mobile Node for the provisioning of a masked prefix IPv6 address. The function takes an encrypted, authenticated Masked Address Configuration Request message and, if no error occurs in processing it, sends the appropriate reply to the Mobile Node. The function has no return on success, but if an error occurs during the processing of the request or reply, the function returns an error indication.

Confidentiality and Authenticity for Masked Address Configuration function

The Confidentiality and Authenticity for Masked Address Configuration function establishes confidentiality and authenticity on the masked address traffic between the Access Router and the Mobile Node. The function takes as one parameter the security association between the Mobile Node and the Access Router. The other parameter is either an encrypted, authenticated Masked Address Configuration Request message from a Mobile Node or a clear text Masked Address Configuration Reply. The function returns either the clear text request or encrypted, authenticated reply, depending on the input parameter, or an error indication if some error occurred during processing.

7.4.9   Next steps in system design

The next step in the system design is to select cryptographic and key distribution algorithms for the security associations between the various functional entities. This decision causes additional functions and interfaces to be added to the architecture. If an AAA-based network access authentication architecture at the link layer is used, key distribution for the Mobile Node to Access Router security association could use shared key cryptography with bootstrapping at network access authentication. Otherwise, a public key algorithm like SEND or possibly using certified keys might be more appropriate. Since the Key Distribution and Masked Address Server and Location Privacy Domain Routers are all in the same routing domain, a shared key algorithm utilizing manually configured, preshared keys is a viable option. In any case, the details of the cryptographic and key distribution algorithms depend heavily on backward compatibility considerations. What algorithms and protocols are currently in use on similar existing interfaces and how could they be leveraged to provide additional functionality at minimal design and implementation cost?

Once these decisions have been made, the interoperable protocols on interfaces CPP1 through CPP3 and on the security association interfaces are designed. Again, backward compatibility with existing protocols is an important consideration. With the exception of the Key Distribution and Masked Address Server, none of the network entities implementing the functions described above is new, and even the Key Distribution and Masked Address Server has functions similar to network entities found in existing systems, such as the AAA server. In the case of routing between Location Privacy Domain Routers, a protocol selection has already been recorded in the architecture, for transporting the encrypted and authenticated partially unmasked address: IPv6 header option. This selection can be left open in the architecture but it is really the only choice given the basic IPv6 protocol format.

7.5   Summary

Privacy in communication between IP nodes is difficult to maintain due to the design of the basic Internet architecture. Internet protocols use identifiers such as the IP address to identify communicating nodes. When identifiers are associated with long-lived traffic flows, they can be used by eavesdropper intermediaries or by end nodes themselves to track what types of activities a particular Internet host undergoes to profile the types of activities particular hosts engage in. Ultimately, such profiling can be traced back to individual users if the identifiers can be mapped to the real-world identity of users owning or using the hosts. Most users have a reasonable expectation that some, if not all, of their online activities remain private to them and possibly the servers and other Internet nodes with which they correspond, unless they are explicitly notified and given the opportunity to release their data.

Location privacy is a specific privacy vulnerability associated with wireless terminals. Identifiers, and specifically the IP address, can be mapped to the geographical location of hosts through some network information, such as the subnet prefix, that is assigned by the network deployment to a particular geographical location. If an attacker can construct a mapping between the geographical location and the identifier, it may be possible for the attacker to track the location of a particular wireless terminal. If the tracking can be done in real time, it may constitute a threat to the user's physical privacy, since the attacker can locate the user and possibly do physical harm. This level of risk is considerably greater than most other network security threats.

Solutions for privacy and location privacy on the network level are relatively few and not very effective. Onion routing allows good privacy, including location privacy, but results in increased latency in network communication, which may mean uncceptable delays for real-time media traffic, such as voice telephone calls. Mobile IP can be deployed and used in a way that provides good location privacy protection. The key factor is to not deploy route optimization in Mobile IPv6, because route optimization exposes the care-of address to the correspondent node, which causes the wireless terminal's location information to leak out. Of course, without route optimization, the wireless terminal does not benefit from reduced route latency, which, again, could impact real-time media traffic with delay constraints. There are a few other deployment issues with maintaining location privacy for Mobile IPv6, and also a few protocol issues that are currently under investigation.

Since the difficulty of maintaining location privacy is a direct result of the basic Internet architecture, architectural solutions can provide a much more powerful and effective solution. The downside of architectural solutions is that they tend to require deep and fundamental changes in network equipment and end hosts. These kinds of changes are expensive and disruptive to existing service, and therefore network operators and their customers are often reluctant to make them. Nevertheless, deep architectural changes are sometimes introduced into existing systems – often after highly publicized attacks where network operators or users suffer serious financial or other harm.

An example of such an architectural change for location privacy is Cryptographically Protected Prefixes (CPP). While CPP is a research scheme and unlikely to actually be deployed, it illustrates how a deep architectural change for location privacy might be introduced into IPv6 networks. CPP decouples the geographic location from the topological location by providing each wireless terminal on a subnet with a masked subnet prefix that is uncorrelated with other masked prefixes on the subnet. Routers within the local routing domain form a location privacy domain, and routing information within the domain is strictly aggregated. All routers at the same level in the hierarchy hold a common level key. A router uses its level key to unmask bits in the subnet prefix allowing standard IPv6 longest prefix mapping against the routing table, in order to find the next hop. Level keys and blocks of addresses with masked subnet prefixes are distributed to the routers from a key distribution and masked address server. A functional architecture for CPP, based on the algorithm, was developed including network interfaces and internal programmatic interfaces. The functional architecture did not include cryptographic and key distribution algorithms for security associations between the functional entities. Next steps in the system design are to define these algorithms and then start standardized protocol design on the interoperable network interfaces. Programmatic interfaces are internal to network entities and therefore the details cannot be standardized, because they will depend on the underlying programming platform.

References

802.1x. (2004). *Port-based Access Control*. IEEE Standard 802.1x–2004. New York: IEEE.

802.11. (1999). *Wireless LAN Medium Access Control (MAC) and Physical Layer (PHY) specifications*. IEEE Standard 802.11–1999. New York: IEEE.

802.11. (2007). *Wireless LAN Medium Access Control (MAC) and Physical Layer (PHY) specifications*. IEEE Standard 802.11–2007. New York: IEEE.

Edney, J. & Arbaugh, W. (2004). *Real 802.11 Security: Wi-Fi Protected Access and 802.11i*. Boston: Addison-Wesley.

Kaufman, C., Perlman, R., & Speciner, M. (2002). *Network Security: PRIVATE Communication in a PUBLIC World*. Upper Saddle River, NJ: Prentice Hall.

Menezes, A., van Oorschot, P., & Vanstone, S. (1997). *Handbook of Applied Cryptography*. Boca Raton, FL: CRC Press.

Qiu, Y., Zhao, F., & Koodli, R. (2007). *Mobile IPv6 Location Privacy Solutions*. Internet Draft. Work in Progress.

RFC 826. Plummer, D. (1982). *An Ethernet Address Resolution Protocol*. Internet Engineering Task Force, Standards Track.

RFC 1256. Deering, S. (1991). *ICMP Router Discovery Messages*. Internet Engineering Task Force, Standards Track.

RFC 1305. Mills, D. (1992). *Network Time Protocol (Version 3) Specification, Implementation and Analysis*. Internet Engineering Task Force.

RFC 1332. McGregor, G. (1992). *The PPP Internet Protocol Control Protocol (IPCP)*. Internet Engineering Task Force, Standards Track.

RFC 1548. Simpson, W. (1993). *The Point-to-Point Protocol (PPP)*. Internet Engineering Task Force, Standards Track.

RFC 1918. Rekhter, Y., Moskowitz, B., Karrenberg, D., de Groot, G., & Lear, E. (1996). *Address Allocation for Private Internets*. Internet Engineering Task Force, Best Current Practice.

RFC 1958. Carpenter, B., editor (1996). *Architectural Principles of the Internet*. Internet Architecture Board.

RFC 2131. Droms, R. (1997). *Dynamic Host Configuration Protocol*. Internet Engineering Task Force, Standards Track.

RFC 2409. Harkins, D. & Carrel, D. (1998). *The Internet Key Exchange (IKE)*. Internet Engineering Task Force, Standards Track.

RFC 2461. Narten, T., Nordmark, E., & Simpson, W. (1998). *Neighbor Discovery for IP version 6 (IPv6)*. Internet Engineering Task Force, Standards Track.

RFC 2462. Thomson, S. & Narten, T. (1998). *IPv6 Stateless Address Autoconfiguration*. Internet Engineering Task Force, Standards Track.

RFC 2794. Perkins, C. & Calhoun, P. (2000). *Mobile IP Network Access Identifier Extension for IPv4*. Internet Engineering Task Force, Standards Track.

RFC 2865. Rigney, C., Livingston, S., Rubens, A., & Simpson, W. (2000). *Remote Authentication Dial In User Service (RADIUS)*. Internet Engineering Task Force, Standards Track.

RFC 3022. Srisuresh, P. & Egevang, K. (2001). *Traditional IP Network Address Translator (Traditional NAT)*. Internet Engineering Task Force, Informational.

RFC 3118. Droms, R. & Arbaugh, W. (2001). *Authentication for DHCP Messages*. Internet Engineering Task Force, Standards Track.

RFC 3280. Housley, R., Polk, W., Ford, W., & Solo, D. (2002). *Internet X.509 Public Key Infrastructure Certificate and Certificate Revocation List (CRL) Profile*. Internet Engineering Task Force, Standards Track.

RFC 3315. Droms, R., Bound. J., Volz. B., Lemon, T., Perkins, C., & Carney, M. (2003). *Dynamic Host Configuration Protocol for IPv6 (DHCPv6)*. Internet Engineering Task Force, Standards Track.

RFC 3344. Perkins, C., editor. (2002). *IP Mobility Support for IPv4*. Internet Engineering Task Force, Standards Track.

RFC 3447. Jonsson, J. & Kaliski, B. (2003). *Public-key Cryptography Standards (PKCS) #1: RSA Cryptography Specifications Version 2.1*. Internet Engineering Task Force, Standards Track.

RFC 3748. Aboba, B., Blunk, L., Vollbrecht, J., Carlson, J., & Levkowetz, H. (2004). *Extensible Authentication Protocol (EAP)*. Internet Engineering Task Force, Standards Track.

RFC 3756. Nikander, P., Kempf, J., & Nordmark, E. (2004). *IPv6 Neighbor Discovery (ND) Trust Models and Threats*. Internet Engineering Task Force, Informational.

RFC 3775. Johnson, D., Perkins, C., & Arkko, J. (2004). *Mobility Support in IPv6*. Internet Engineering Task Force, Standards Track.

RFC 3957. Perkins, C., & Calhoun, P. (2005). *Authentication, Authorization, and Accounting (AAA) Registration Keys for Mobile IPv4*. Internet Engineering Task Force, Standards Track.

RFC 3971. Arkko, J., Kempf, J., Zill, B., & Nikander. P. (2005). *SEcure Neighbor Discovery (SEND)*. Internet Engineering Task Force, Standards Track.

RFC 3972. Aura, T. (2005). *Cryptographically Generated Addresses (CGA)*. Internet Engineering Task Force, Standards Track.

RFC 4187. Arkko, J. & Haverinen, H. (2007). *Extensible Authentication Protocol Method for 3rd Generation Authentication and Key Agreement (EAP-AKA)*. Internet Engineering Task Force, Informational.

RFC 4193. Hinden, R. & Haberman, B. (2005). *Unique Local IPv6 Unicast Addresses*. Internet Engineering Task Force, Standards Track.

RFC 4282. Aboba, B., Beadles, M., Arkko, J., & Eronen, P. (2005). *The Network Access Identifier*. Internet Engineering Task Force, Standards Track.

RFC 4291. Deering, S. & Hinden, R. (2006). *IP Version 6 Addressing Architecture*. Internet Engineering Task Force, Standard Track.

RFC 4301. Kent, S. & Seo, K. (2005). *Security Architecture for the Internet Protocol*. Internet Engineering Task Force, Standards Track.

RFC 4302. Kent, S. (2005). *IP Authentication Header*. Internet Engineering Task Force, Standards Track.

RFC 4303. Kent, S. (2005). *IP Encapsulating Security Payload (ESP)*. Internet Engineering Task Force, Standards Track.

RFC 4306. Kaufman, C. (2005). *Internet Key Exchange (IKEv2) Protocol*. Internet Engineering Task Force, Standards Track.

RFC 4346. Dierks, T. & Rescoria, E. (2006). *The Transport Layer Security (TLS) Protocol Version 1.1*. Internet Engineering Task Force, Standards Track.

RFC 4449. Perkins, C. (2006). *Securing Mobile IPv6 Route Optimization Using a Static Shared Key*. Internet Engineering Task Force, Standards Track.

RFC 4632. Fuller, V. & Li, T. (2006). *Classless Inter-domain Routing (CIDR): The Internet Address Assignment and Aggregation Plan*. Internet Engineering Task Force, Best Current Practice.

RFC 4721. Perkins, C., Calhoun, P., & Bharatia, J. (2007). *Mobile IPv4 Challenge/Response Extensions (Revised)*. Internet Engineering Task Force, Standards Track.

RFC 4861. Narten, T., Nordmark, E., Simpson, W., & Soliman, H. (2007). *Neighbor Discovery for IP version 6 (IPv6)*. Internet Engineering Task Force, Standards Track.

RFC 4862. Thomson, S., Narten, T., & Jinmei, T. (2007). *IPv6 Stateless Address Autoconfiguration*. Internet Engineering Task Force, Standards Track.

RFC 4866. Arkko, J. *Enhanced Route Optimization for Mobile IPv6*. Internet Engineering Task Force, Standards Track.

RFC 4882. Koodli, R. (2007). *IP Address Location Privacy and Mobile IPv6: Problem Statement*. Internet Engineering Task Force, Informational.

RFC 4941. Narten, T., Draves, R., & Krishnan, S. (2007). *Privacy Extensions for Stateless Address Autoconfiguration in IPv6*. Internet Engineering Task Force, Standards Track.

RFC 4962. Housley, R. & Aboba, B. (2007). *Guidance for Authentication, Authorization, and Accounting (AAA) Key Management*, Internet Engineering Task Force, Best Current Practices.

Syverson, P., Goldschlag, D., & Reed, M. (1998). Anonymous Connections and Onion Routing. *Proceedings of the 18th Annual Symposium on Security and Privacy*, Oakland, CA: IEEE CS Press May 1997. pp. 44–54.

Trostle, J., Matsuoka, H., bin Tarq, M., Kempf, J., Kawahara, T., & Jain, R. (2005). Cryptographically Protected Prefixes for Location Privacy in IPv6. *Privacy Enhancing Technologies*. Lecture Notes in Computer Science, 3424. Berlin: Springer.

Wikipedia. (2008a). *OSI Model*. http://en.wikipedia.org/wiki/OSI_model.

Wikipedia. (2008b). *RSA*, http://en.wikipedia.org/wiki/Rsa

Wikipedia. (2008c). *MAC Address*. http://en.wikipedia.org/wiki/MAC_address

Wikipedia. (2008d). *International Mobile Subscriber Identity*. http://en.wikipedia.org/wiki/International_Mobile_Subscriber_Identity.

Index

Note: Figures and Tables are indicated by *italic* page numbers.